International Perspectives in Geography

AJG Library

Volume 22

Aim and Scope

The AJG Library is published by Springer under the auspices of the Association of Japanese Geographers. This is a scholarly series of international standing. Given the multidisciplinary nature of geography, the objective of the series is to provide an invaluable source of information not only for geographers, but also for students, researchers, teachers, administrators, and professionals outside the discipline. Strong emphasis is placed on the theoretical and empirical understanding of the changing relationships between nature and human activities. The overall aim of the series is to provide readers throughout the world with stimulating and up-to-date scientific outcomes mainly by Japanese and other Asian geographers. Thus, an "Asian" flavor different from the Western way of thinking may be reflected in this series. The AJG Library will be available both in print and online via SpringerLink.

About the AJG

The Association of Japanese Geographers (AJG), founded in 1925, is one of the largest and leading organizations on geographical research in Asia and the Pacific Rim today, with around 3000 members. AJG is devoted to promoting research on various aspects of human and physical geography and contributing to academic development through exchanges of information and knowledge with relevant internal and external academic communities. Members are tackling contemporary issues such as global warming, air/water pollution, natural disasters, rapid urbanization, irregular land-use changes, and regional disparities through comprehensive investigation into the earth and its people. In addition, to make the next generation aware of these academic achievements, the members are engaged in teaching and outreach activities of spreading geographical awareness. With the recent developments and much improved international linkages, AJG launches the publication of the AJG Library series in 2012.

Satoshi Yokoyama · Takahito Niwa · Hongwei Jiang
Editors

Population Dynamics and Livelihood Changes of Small-Scale Societies in Laos

Springer

Editors
Satoshi Yokoyama
Nagoya University
Nagoya, Aichi, Japan

Takahito Niwa
Kanazawa University
Kanazawa, Ishikawa, Japan

Hongwei Jiang
National Institute for Environmental Studies
Tsukuba, Ibaraki, Japan

ISSN 2197-7798 ISSN 2197-7801 (electronic)
International Perspectives in Geography
ISBN 978-981-96-2955-8 ISBN 978-981-96-2956-5 (eBook)
https://doi.org/10.1007/978-981-96-2956-5

This Springer imprint is published by the registered company Springer Nature Singapore Pte Ltd.
The registered company address is: 152 Beach Road, #21-01/04 Gateway East, Singapore 189721, Singapore

Preface

It is undisputed that understanding the population dynamics of small-scale communities in emerging and developing countries, which account for a large proportion of the global population, has attracted the attention of various academic disciplines concerned with population. However, the population dynamics of small-scale communities in these countries has not yet been studied in detail. In this study, we focus on small-scale communities in Laos, where the importance of money is rapidly increasing in the transition economy and some small-scale communities are changing their livelihood structure to obtain cash, while many small-scale communities are still completely self-sufficient. The objective of this study was to identify livelihood strategies at the household level in Laos, where many small-scale communities continue to rely on subsistence production.

This book is the result of the following two Japan Society for the Promotion of Science (JSPS) projects. The first project built on the results of research on the relationship between population dynamics and livelihoods and the internal factors of population reproduction in rural Laos. The second project delved deeper into migration, livelihoods, food acquisition, and family planning to elucidate the demographic mechanisms of small-scale communities in contemporary Laos.

1. *Population Dynamics, Reproduction and Livelihood Changes in Small-scale Communities of Laos* (FY2013–2016, Grant-in-Aid for Scientific Research (A), project No. 25257004, Principal Investigator: Satoshi YOKOYAMA)
2. *The Mechanism of Population Dynamics in Small-scale Communities of present Laos* (FY2017–2020, Grant-in-Aid for Scientific Research (A), project No. 17H01633, Principal Investigator: Satoshi YOKOYAMA)

Although this project focused on only three areas in Laos, a landlocked country in mainland Southeast Asia, the results of the study differed from the macroscopic population-food relationship that has been discussed on a global scale. This allowed us to present a realistic picture of the people's way of life. The study of population dynamics in one village (hamlet) over three generations, and its relationship to population dynamics and livelihood changes, provided extremely valuable information for discussing future trends of small-scale communities.

In preparing this book, we have been greatly assisted by many people in Laos, including staff and researchers from the National Agriculture and Forestry Research Institute, the Lao Tropical and Public Health Institute, Savannakhet Provincial Health Department, Xepon District Health Office in Savannakhet Province, Ngoi District Health Office in Luang Prabang Province, and Ngoi District Agriculture and Forestry Office in Luang Prabang Province.

Finally, we would like to express our gratitude to the members of the Editorial Board of International Perspectives in Geography: AJG Library, in AJG, for their invaluable contribution in reviewing the contents of this book.

Nagoya, Japan — Satoshi Yokoyama
Kanazawa, Japan — Takahito Niwa
Tsukuba, Japan — Hongwei Jiang

Contents

Editors and Contributors

About the Editors

Dr. Satoshi Yokoyama is a Professor in Department of Geography, Graduate School of Environmental Studies, Nagoya University. His research interests lie in the field of human–nature interactions, with a particular emphasis on land use change, natural resource use, and livelihood change. His most recent research interests include traditional fermented foods in the Himalayas and Southeast Asia. He has authored or edited 11 books, and published over a hundred academic book chapters and research articles. He is a member of the Science Council of Japan (SCJ) since October 2020 and a program officer of the Research Center for Science Systems, Japan Society for the Promotion of Science (JSPS) since April 2023. Additionally, he held the position of a Visiting Scholar at East-West Center, USA. from 2005 to 2006; a Visiting Associate Professor at the Research Institute for Humanity and Nature in 2011; and a Visiting Professor at Center for Southeast Asian Studies, Kyoto University from 2013 to 2014.

Dr. Takahito Niwa is an Associate Professor in Collage of Human and Social Sciences, Kanazawa University. He received his Ph.D. in Science at the Graduate School of Science, Tohoku University in 2012. His specialties are economic geography and population geography. His research focuses on the spatial patterns of population migration and adaption of workers in local labor markets around Southeast Asia. He has continued to research internal labor migration in Thailand, transnational migration between Laos and Thailand, lifestyle migration among Japanese youth, and retirement migration among Japanese elderly, based on empirical case studies. He has recently published his edited book on "Current global human resources" (in Japanese) in 2018.

Dr. Hongwei Jiang is a Resarch Fellow in Health and Environmental Risk Division, National Institute for Environmental Studies, Japan. His research interests lies in the field of human ecology with emphasis on subsistence transition and human health, spatio-temporal analysis of human activities, ecohealth. His recent research interests include land use change and emerging infectious disease (EID) risk, and he published an article entitled "Factors Contributing to the Pre-Elimination of Malaria from Hainan Island, China, 1986–2009" on the American Journal of Tropical Medicine and Hygiene. Additionally, he is a Visiting Assistant professor at the Research Institute of Humanity and Nature.

Contributors

Hongwei Jiang Graduate School of Humanities, Osaka University, Toyonaka, Japan;
National Institute for Environmental Studies, Tsukuba, Japan

Yoshie Moriki College of Liberal Arts, International Christian University, Mitaka, Tokyo, Japan

Futoshi Nishimoto Embassy of Japan in Laos, Vientiane, Laos

Takahito Niwa College of Human and Social Sciences, Kanazawa University, Kanazawa, Japan

Ren'ya Sato Graduate School of Humanities, Osaka University, Toyonaka, Japan

Chihiro Shirakawa Graduate School of Human Sciences, Osaka University, Suita, Osaka, Japan

Shinichi Takahashi Kobe University, Kobe, Japan

Satoshi Yokoyama Graduate School of Environmental Studies, Nagoya University, Nagoya, Japan

List of Figures

Population Growth and Generational Response to Livelihoods in a Rain-fed Paddy Farming Village of Central Laos

Transnational Labor Migration from Rural Village in Laos to Bangkok, Thailand

Growth in the Use of Modern Contraceptive Methods and the Related Social Context in a Central Laotian Farming Village

Population Dynamics and Paddy Holdings in Remote Rural Villages of Northern Laos

Rural-Rural Migration in a Rural Village, Luang Phabang Province, Northern Laos

Rural–Urban Migration in Northern Laos

The Implications for the Study on Population Dynamics and Livelihood Changes in Small-Scale Communities in Laos

List of Tables

Food Procurement Strategy of Swidden Agriculturalists in Central Laos

Temporal-spatial Distribution of Swidden Agriculturists' Daily Activities in Central Laos

Population Growth and Generational Response to Livelihoods in a Rain-fed Paddy Farming Village of Central Laos

Transnational Labor Migration from Rural Village in Laos to Bangkok, Thailand

Growth in the Use of Modern Contraceptive Methods and the Related Social Context in a Central Laotian Farming Village

Population Dynamics and Paddy Holdings in Remote Rural Villages of Northern Laos

Rural-Rural Migration in a Rural Village, Luang Phabang Province, Northern Laos

Rural–Urban Migration in Northern Laos

Aspects of Modern Contraceptive Use Among Women in Rural Northern Laos

Family Planning Methods and Fertility Level in a Northern Remote Village of Laos: From Viewpoints of Women's Autonomy and International Health Cooperation

The Implications for the Study on Population Dynamics and Livelihood Changes in Small-Scale Communities in Laos

Prologue

Population Dynamics and Livelihood Change of Small-Scale Society in Laos

Satoshi Yokoyama

Abstract This prologue reviews previous studies on the relationship between population dynamics and livelihood change, and provides background on the research and an overview of this book. To date, discussions of the relationship between population and livelihoods, particularly food supply, have been conducted from a global perspective. It has been argued that even as population increases, innovations in agricultural technology will prevent food shortages. However, it is difficult to apply discussions at the macro-global level to the micro-village level because people's real lives are led in small-scale societies. Each small-scale social group has its own unique population dynamics, reproduction, and livelihood structure, and the complex interrelationships among them need to be elucidated. Despite the fact that small-scale social groups in emerging and developing countries constitute the majority of the world's population, it has been difficult to understand population dynamics in these countries where statistics are not well developed. In this book, we discuss the population dynamics of small-scale social groups in Laos, which still have subsistence livelihood structures, by reconstructing the population, reproduction, and livelihoods of the past three generations and discussing the actual village dynamics where people live.

Keywords Population dynamics · Reproduction · Livelihood change · Paddy farming · Swidden · Migration · Laos

1 Study Background and Objectives

The interrelationship between population change, reproduction, and livelihood change is an extremely important theme of study when considering the foundations of human survival in an era of globalization. To date, numerous global-scale studies

This chapter is a revised version of Yokoyama (2023) with the addition of a book summary.

S. Yokoyama (✉)
Graduate School of Environmental Studies, Nagoya University, Nagoya, Japan
e-mail: yokoyama.satoshi.i4@f.mail.nagoya-u.ac.jp

S. Yokoyama et al. (eds.), *Population Dynamics and Livelihood Changes of Small-Scale Societies in Laos*, International Perspectives in Geography 22,
https://doi.org/10.1007/978-981-96-2956-5_1

have been made with regard to issues such as population and food in connection with the foundations for human survival. The oldest of these may be the population theory of Thomas Robert Malthus, which predicted that the supply of land resources (food) would not be able to keep up with population growth, resulting in "checks" that would limit population growth, such as war, disease, and famine (Malthus 1872). However, the "Malthusian trap," that land resource limitations would limit economic growth and population growth provided to be an unfounded fear. In reality, as population pressure increases, agricultural intensification characterized by shorter fallow periods, proceeds. It was Ester Boserup who explained the concept (Boserup 1965). However, objections have arisen in response to Boserup's theory that along with agricultural intensification, while production volume does increase, the labor input also increases, which means that farmers shift from high labor productivity to swidden farming to high land productivity to paddy fields (e.g., Brookfield 1972). On the other hand, Ellis et al. (2013), who analyzed land use and population change on a global scale over the past 300 years, concluded that human history of overcoming difficulties through various regime shifts despite the exponential population growth actually supports the validity of Boserup's theory.

Certainly, on a global scale, it cannot be denied that the development of agricultural technology has made it possible to fully support the ever-growing population by providing sufficient food acquisition. Humankind has progressed toward agricultural intensification, and grain production has doubled in the last 40 years. Mathematically, a sufficient amount of food is being distributed to the global population. However, it is also true that there are still many undernourished people, which means that discussing the relationship between population fluctuations and food supply through global statistics is nothing more than an empty theory.

According to Hazell and Rahman (2014), 450 million farmers around the world are estimated to be smallholders with less than 2 ha of farmland. Yet, in the small-scale societies where those smallholders live, the relationship between population dynamics and food acquisition is determined by norms that differ from that of the maximization of economic profit. As an example, Geertz (1963) used the phrase "shared poverty" to describe the approach to work opportunities and income distribution among farmers in Java, Indonesia, under Dutch colonial rule. Scott (1976) used the phrase "moral economy" to refer to the behavioral principles and value norms of farmers in colonial Vietnam and Myanmar related to risk avoidance and the prioritization of security in uncertain agricultural production.

The actual activities of people are carried out according to various local norms, with small-scale social groups as the basic unit. In recent years, with the addition of informatization and modernization associated with globalization, and the further penetration of the concepts of family planning and public health, food production can no longer be regarded as the only factor defining the population. Therefore, research must explore the relationship between the population and the economy, society, culture, disease, and hygiene, as well as their impact on population dynamics.

However, although research has been accumulated in developed countries with a population registration system and a national census, research on emerging and developing countries, where the development of various statistics is lagging behind, is very limited. A representative study that clarified the relationship between population dynamics and livelihoods of small-scale social groups through field research without relying on official statistics was conducted by the Center for Southeast Asian Studies, Kyoto University, in the 1980s and 1990s in Dong Daeng village near Khon Kaen, a major city in northeastern Thailand. This long-term village survey by humanities, social science and natural science researchers was unprecedented; the Dong Daeng village survey produced significant findings related to population dynamics and livelihoods, as well as social transformation (Kuchiba 1990; Fukui 1993). As a project, the Dong Daeng village study ended in the 1990s, but research results related to agroecology by individual researchers continued after the 2000s (Watanabe et al. 2007; Miyagawa et al. 2006, 2011, 2019).

Studies of population dynamics, livelihood change, and social transformation in northeastern Thailand, where Dong Daeng village is located, have shown that there are increasing opportunities for work in the capital city of Bangkok, work in other countries, and off-farm work due to the expansion of nearby cities, as well as an increase in extensive agriculture due to population out-migration (Rigg and Salamanca 2011; Rigg et al. 2012, 2018; Rigg 2013; Le Mare et al. 2015; Shirai and Rambo 2017). However, there are no detailed household-level studies, such as those conducted in Dong Daeng village, of who in the household has gone where, who is employed in what occupations, and what strategies are employed at the household level to live.

Meanwhile, in Laos, a series of policies have been implemented with the aim of protecting forests and creating stable employment opportunities (Évrard and Baird 2017; Molland 2017). Consequently, since the 1990s, the practice of subsistence swidden agriculture has diminished in remote mountainous areas, and migration to lowlands with road access has been promoted. Furthermore, in the central and southern regions along the Mekong River, the number of people engaging transnational migration work to Thailand from has increased (Manivong et al. 2014; Phouxay 2017). In recent years, the birthrate in Laos has also been declined, as it has in other Southeast Asian countries (Yeung 2022). There is a notable scarcity of agricultural labor (Sacklokham et al. 2020). However, in Laos, where statistical data are not well developed, obtaining household-level data requires significant effort, and thus no research has been conducted to clarify the relationship between population dynamics, livelihood changes, and social transformation.

In a country where statistics are not yet available, it takes a lot of effort to obtain data at the household level, and thus it has been possible only to infer the causes of population change from fragmentary information on changes in livelihoods.

Various academic disciplines that concerned with population are interested in the dynamics of small-scale social groups in emerging and developing countries, which make up the world's population. However, the population dynamics of small-scale societies in these countries have not been studied in detail. In Laos, many small-scale social groups are still engaged in subsistence activities, but money is rapidly becoming more important due to the transitioning economy. Their livelihood structure is changing from subsistence to cash income activities. Therefore, a research project was conducted to select small-scale social groups engaged in subsistence livelihood activities and to clarify their livelihood strategies at the household level in Laos.[1] This book is compiled as a result of the research project.

2 Study Areas and Methods

2.1 Study Areas

The study areas in this book are the small-scale social groups in the landlocked country of Laos in Southeast Asia. Three areas (three villages and one hamlet) are covered (Fig. 1), including AN hamlet[2] of the Mankhong people in Savannakhet province, where people are engaged in subsistence swidden upland rice farming; KH village of the ethnic Lao people in Savannakhet province, central Laos, where people are engaged in subsistence rain-fed rice farming; and NK and HB villages, with a mix of ethnic Lao and Khmu people in Luang Phabang province, northern Laos, where people are engaged in subsistence rain-fed rice farming and swidden upland rice farming.

AN hamlet of the Mankhong (Fig. 2a), located in a basin in central Laos far from the Thai border, the mother tongue of the Mankhong people, the Mon-Khmer linguistic group, is not Lao; thus, with a linguistic barrier as well as geographical separation from Thailand, none of the residents have migrated to Thailand for work. In addition, the livelihood structures are very different from those of the ethnic Lao villages, whose residents are engaged in paddy rice farming in the lowlands, as they are engaged in swidden upland rice farming and hunting/gathering.

On the other hand, in KH village (Fig. 2b), a rain-fed rice farming village, is located in the lowlands of central Laos, which has seen a rapid increase in transnational migrants to Thailand since the 1990s due to rising wages caused by economic growth in Thailand. Since the Thai and Lao languages are similar, which allows ethnic Lao to understand Thai, there are almost no language barriers to working in Thailand. Geographically, Thailand is accessible just across the Mekong River, and many people illegally across the border into Thailand (Phouxay 2017).

NK village (Fig. 2c) and HB village (Fig. 2d) in the basin of northern Laos were originally ethnic Lao villages, but due to resettlement by the Lao government policy,

---

[1] Grants-in-Aid for Scientific Research, Japan Society for the Promotion of Science (JSPS) were received to assist with the implementation of this study. They are, "Population dynamics, reproduction and livelihood changes in small-scale communities of Laos" (FY2013-2016, Grant-in-Aid for Scientific Research (A), project number 25257004, Principal Investigator: Satoshi YOKOYAMA); and "The mechanism of population dynamics in small-scale communities of present Laos" (FY2017-2020, Grant-in-Aid for Scientific Research (A), project number 17H01633, Principal Investigator: Satoshi YOKOYAMA).

[2] The AN hamlet merged with the neighboring AG hamlet to be known administratively as AG village. However, the AG hamlet was outside the scope of the study; hence, the papers here continue to use the term "AN hamlet."

Fig. 1 Study area

a. AN hamlet (February, 2020)
b. KH village (December, 2014)
c. NK village (August, 2017)
d. HB village (June, 2017)

Fig. 2 Landscape of the study area. *Source* Photo by author

the Khmu people have been moving from the mountains to the basin since the mid-1970s, and the two groups have become mixed. The main livelihood is rice cultivation based on traditional irrigated paddy fields, but only a limited number of households own paddy fields, and many of the Khmu resettled from the mountains are engaged in swidden upland rice farming. There are no roads accessible by car, there is no grid electricity, and each household generates its own electricity using hydroelectric power from the stream near the village. Furthermore, cell phone reception is extremely weak in this region, where infrastructure is severely lacking. Since the mid-1990s, an increasing number of people from both NK and HB villages have migrated to work in urban areas of Laos.

2.2 *Research Methods*

In these three regions of Laos, with their different characteristics, in order to clarify the dynamics of small-scale social groups in this research project, surveys were conducted with all households, collecting data on population dynamics, reproduction, and livelihoods as far back as possible.

In KH, NK, and HB villages of the ethnic Lao, it is possible to estimate the population going back to the 1910s and 1920s. These population data were obtained by tracing the family trees of current residents. Where possible, data was also collected for households that had moved away and households that had died out, but in some cases they had already disappeared from memory and it was not possible to recover the data. Therefore, accurate data that could withstand analysis were obtained for about three generations, that is, about 50–60 years, from residents in their eighties with reliable memories and who know the village well. While some article in this book discuss the population of the 1920s, their objective is to investigate the inheritance and distribution of the paddy fields owned by the current households. Descriptions of population dynamics before the 1970s are given only for reference.

As for the reproduction data, it is essential to know the number of births by age, including stillbirths, to obtain fertility and mortality rates, but these data are obtained within the knowledge of the current residents. Therefore, data from three generations ago are estimates.

The population data for KH, NK, and HB villages were entered into the genealogical application "Gramps"[3] and shared among the research team members. The paddy field data for these three villages were entered into GIS using high-resolution satellite imagery, while swidden field data for the AN hamlet were surveyed using GPS and entered into GIS. These GIS attribute data were also shared among the team. All articles in this book provide analyses using these shared data. Regarding the AN hamlet, population dynamics and reproduction data were also acquired, but the elderly villagers were not sure even of their own age, therefore the analysis was conducted with more focus on livelihood than on population dynamics.

3 Summary of the Book

Since this study covered three study areas in Laos that differ in both geographic setting and livelihood structure, this book is divided into three parts to discuss each of these areas.

Part I contains two articles (Sato et al. 2025; Jiang et al. 2025) that discuss AN hamlet, where people are engaged in swidden farming, fishing, livestock raising, hunting, and gathering to maintain self-sufficiency in eating habits. In AN hamlet, there is no movement of internal or transnational migration, and traditional swidden farming continues as the main livelihood, although the influence of commercialization can be seen in recent years.

Part II contains three articles (Takahashi et al. 2025; Niwa et al. 2025b; Shirakawa et al. 2025) that discuss KH village, where the international migrant population to Thailand has been increasing since the 1990s and recently accounted for more than 35% of the permanent population. Until the 1990s, the population growth rate of KH village was very high, more than 4% per year. However, with the increase in the number of migrant workers, family planning promotion activities began and the birth

[3] Open-source genealogical software that creates databases from relationships, such as parents, siblings, and adopted children, can be downloaded from the following link: https://gramps-project.org/blog/.

rate declined. In addition, although rain-fed farming is the main source of livelihood, in recent years there has been no more space to reclaim paddy fields. Then, as income has increased due to migrant work, the method of obtaining paddy fields has shifted from inheritance to purchase.

Part III contains five articles (Yokoyama et al. 2025; Niwa et al. 2025a; Niwa and Nishimoto 2025; Shirakawa and Moriki 2025; Moriki and Shirakawa 2025) that discuss NK and HB villages, which have received many migrants from mountain villages since the 1970s, as government policies have encouraged migration from mountainous areas to lowlands. In the basin where NK and HB villages are located, there was limited land for new paddy fields, and therefore many residents who migrated from the mountainous areas could not obtain paddy fields and continued subsistence swidden farming as they had practiced in the mountainous areas. Purchasing paddy fields from residents who had left the village was the only way to obtain paddy fields, and many residents began to wait for such opportunities and went to work in urban areas to earn money to purchase paddy fields.

A summary of each article is given below.

3.1 Part I: Livelihoods in a Swidden Agriculture Community

Sato et al. (2025) obtained annual eating habit data from food diaries and analyzed household food procurement strategies. The results showed that children have a positive effect on household living expenses from the age of 10, and thus, the growth of children makes an important contribution to family finances.

Jiang et al. (2025) analyzed the time-allocation of residents' daily life, spatial information of their daily activities using GPS loggers, and energy consumption using accelerometers, separately for the wet and dry seasons. The results showed that when considering the temporal-spatial patterns of daily life in swidden agriculturists where development is progressing rapidly, in addition to the main livelihood, which is rice cultivation through swidden farming, there is a need to grasp livelihoods as a whole, with attention to gathering activities of commercialized natural resources as secondary activities.

3.2 Part II: Population Dynamics and Livelihoods in a Paddy Farming Community in the Plain

Takahashi et al. (2025) analyzed generational differences in paddy field acquisition processes among residents of KH village, which has a long-term population increase of over 2% per year, with a focus on rice cultivation for rice acquisition, which is a staple food. The results showed that along with the population increased, the means of acquiring paddy fields had shifted from reclamation to inheritance and then to acquisition through purchase, thus clarifying the connection with the population stagnation in recent years due to transnational labor migration and the use of family planning.

Niwa et al. (2025b) examines how village-level network connections support transnational labor migration from Laos to Bangkok, Thailand. With the development of ICT, villagers can imagine what life and work are like in Bangkok before they leave. Furthermore, in recent years, Thai employers have become more compliant, migrants have been granted residential permits and work permits, lowering the barriers to migration to Bangkok and the systemic barriers to transnational labor migration, so the number of migrant workers continues to increase. Nevertheless, working conditions have not sufficiently improved, and migrant workers found to make use of local KH village networks to avoid the risk of ill-treatment.

Shirakawa et al. (2025) discusses the process and characteristics of the diffusion of the use of modern contraceptive methods among KH villagers and the social background that led to its diffusion. The diffusion of modern contraceptive use in KH villages since the late 1990s has led to a decline in the fertility rate and the average number of births per woman. The Lao government's family planning program, which began in 1996, has popularized the use of modern contraceptive methods and made them available to rural women. However, many women had already begun to use them before then, and transnational migration to Thailand can be pointed to as a possible source of this trend.

3.3 Part III: Population Dynamics and Livelihoods in Both Paddy and Swidden Farming Communities in the Basin

Yokoyama et al. (2025) focuses on NK and HB villages, analyzing changes in population dynamics and paddy field holding over three generations. In both villages, there has been an increase in Khmu people from the mountains since the 1970s, but most of them have limited opportunities to own paddy fields in the new location, resulting in the inevitable continuation of subsistence swidden farming just as they had done when living in the mountains.

Niwa et al. (2025a) clarifies the characteristics of households moving into and out of HB village in the northern basin of Laos. In particular, it discusses the accessibility of farmland and the networks of kinship and community ties of these households in terms of inter-rural migration due to migration policies and migration for marriage.

Niwa and Nishimoto (2025) conducted interviews with 42 households in HB village and 22 migrants from the village to Vientiane to examine the characteristics of the rural–urban population mobility. The results revealed that rural–urban migrants initially enter the urban labor market seeking low-skilled jobs, then change jobs to improve their working conditions and living standards, and finally settle in the urban area and leave the village permanently.

Shirakawa and Moriki (2025) clarifies the use of modern contraceptive methods and investigates their characteristics based on interviews using a questionnaire with 80 adult female residents in NK village. The results show that almost half of the women have used modern contraceptive methods, and especially, well over half of the young women in their twenties and thirties use these methods based on information from medical personnel. Modern contraceptive methods are used by middle-aged women in their forties and fifties, who already have a large number of children, to avoid having more children, while women in their twenties and thirties use these methods for planned birth control while they are still in the early stage of life, showing the difference in use between generations.

Moriki and Shirakawa (2025) analyzed generational differences in fertility behavior and contraceptive use among married women aged 18 and older in NK village. The results showed that young adults in their twenties and thirties had statistically significantly lower numbers of births and surviving children than older adults in their forties and older. The fertility level in NK village was found to be at a level that should raise concerns about further decline. By comparing with the Shirakawa et al. (2025), which discusses the case of a rural village in central Laos, it was pointed out that the case of NK village in a rural village in the Northern Laos also follows the trend of the declining birth rate in Laos as a whole.

4 Conclusion

This book analyzes data on population and livelihoods in terms of the interrelationships among factors to elucidate which variables affect the dynamics of small-scale social groups. Although the study was conducted in only three regions of Laos, a landlocked country in mainland Southeast Asia, the results differ from the macroscopic relationship between population and food that has been discussed on a global scale and provide a realistic picture of the way people live. This study, which examines the population dynamics of a village over three generations and relates it to changes in livelihoods, provides extremely valuable information for discussing the future trends of small-scale social groups, as do the results of a research project in Dong Daeng village conducted primarily by Kyoto University.

The data on population dynamics and livelihoods of three generations in Laos cover important events such as the upheaval caused by the Vietnam War in the 1970s, the change in political system from monarchy to socialism, and economic liberalization since the late 1980s. Therefore, this study can be placed as a valuable research material as a regional geography of Laos that deals with people's responses to political, economic, and social changes.

Acknowledgements This study was supported from JSPS Grants-in-Aid JP25257004 and JP17H01633. This research project was conducted with the cooperation of the residents of KH village, AN hamlet, NK village, and HB village, and the staff of the counterparts at Xepon and Songkhone Districts in Savannakhet Province, Ngoi District in Luang Phabang Province, the Ministry of Agriculture and Forestry, the National Agriculture and Forestry Research Institute of Laos, the Ministry of Health, and the Tropical and Public Health Institute, to whom the authors are most grateful. Finally, the late Professor Satoshi NAKAGAWA (Population Geography, Saitama University, Japan) participated in this research project from 2013 to 2017; and contributed greatly to the understanding of population dynamics in KH village. However, the members of the research project were deeply saddened by his passing in 2019. This book is dedicated to late Prof. Satoshi Nakagawa.

References

Boserup E (1965) The conditions of agricultural growth. Allen & Unwin, London

Brookfield HC (1972) Intensification and disintensification in Pacific agriculture: a theoretical approach. Pac Viewp 13:30–48. https://doi.org/10.1111/apv.131003

Ellis EC, Kaplan JO, Fuller DQ, Vavrus S, Goldewijk KK, Verburg PH (2013) Used planet: a global history. Proc Natl Acad Sci 110:7978–7985. https://doi.org/10.1073/pnas.1217241110

Évrard O, Baird IG (2017) The political ecology of upland/lowland relationships in Laos since 1975. In: Bouté V, Pholsena V (eds) Changing lives in Laos: society, politics and culture in a post-socialist state. NUS Press, Singapore, pp 165–191

Fukui H (1993) Food and population in a Northeast Thai village. University of Hawaii Press, Honolulu

Geertz C (1963) Agricultural involution: the process of ecological change in Indonesia. University of California Press, Berkeley

Hazell P, Rahman A (eds) (2014) New directions for smallholder agriculture. Oxford University Press, Oxford

Jiang H, Sato R, Yokoyama S, Nishimoto F (2025) Temporal-spatial distribution of swidden agriculturists' daily activities in central Laos. In: Yokoyama S, Niwa T, Jiang H (eds) Population dynamics and livelihood changes of small-scale societies in Laos. Springer, Singapore, pp. 45–69. https://doi.org/10.1007/978-981-96-2956-5_3

Kuchiba M (ed) (1990) Don Daeng mura no dentōkōzō to sono henyō (Traditional structure of Dong Daeng village and its transformation). Sobunsha, Tokyo (in Japanese)

Le Mare A, Promphaking B, Rigg J (2015) Returning home: the middle-income trap and gendered norms in Thailand. J Int Dev 27:285–306. https://doi.org/10.1002/jid.3064

Malthus TR (1872) An essay on the principle of population. Reeves and Turner, London

Manivong V, Cramb R, Newby J (2014) Rice and remittances: crop intensification versus labour migration in southern Laos. Hum Ecol 42:367–379. https://doi.org/10.1007/s10745-014-9656-6

Miyagawa S, Matsuda A, Watanabe M (2019) Transition of home garden cultivation in a rainfed rice-growing village in northeast Thailand from the 1980s to the 2010s. Trop Agric Dev 63:167–175. https://doi.org/10.11248/jsta.63.167

Miyagawa S, Tsuji T, Matsuyama K, Hattori Y, Seko M (2011) Combined effects of recently introduced cultivation practices on rice production in rain-fed paddy fields of northeast Thailand. Trop Agric Dev 55:97–102. https://doi.org/10.11248/jsta.55.97

Miyagawa S, Tsuji T, Watanabe K, Hoshikawa K (2006) Long-term and spatial evaluation of rice crop performance of rain-fed paddy fields in a village of Northeast Thailand. Tropics 15:39–49. https://doi.org/10.3759/tropics.15.39

Molland S (2017) Migration and mobility in Laos. In: Bouté V, Pholsena V (eds) Changing lives in Laos: society, politics and culture in a post-socialist state. NUS Press, Singapore, pp 327–349

Moriki Y, Shirakawa C (2025) Family planning methods and fertility level in a northern remote village of Laos: from viewpoints of women's autonomy and international health cooperation. In: Yokoyama S, Niwa T, Jiang H (eds) Population dynamics and livelihood changes of small-scale societies in Laos. Springer, Singapore, pp 223–234. https://doi.org/10.1007/978-981-96-2956-5_11

Niwa T, Nishimoto F (2025) Rural-urban migration in northern Laos: a case study of HB village in Luang Phabang Province. In: Yokoyama S, Niwa T, Jiang H (eds) Population dynamics and livelihood changes of small-scale societies in Laos. Springer, Singapore, pp 191–206. https://doi.org/10.1007/978-981-96-2956-5_9

Niwa T, Nishimoto F, Takahashi S, Yokoyama S (2025a) Rural-rural migration in a rural village, Luang Phabang Province, northern Laos: Focusing both on immigration and emigration in HB village. In: Yokoyama S, Niwa T, Jiang H (eds) Population dynamics and livelihood changes of small-scale societies in Laos. Springer, Singapore, pp 171–189. https://doi.org/10.1007/978-981-96-2956-5_8

Niwa T, Takahashi S, Nishimoto F (2025b) Transnational labor migration from rural village in Laos to Bangkok, Thailand: Focusing on relationship with the home village. In: Yokoyama S, Niwa T, Jiang H (eds) Population dynamics and livelihood changes of small-scale societies in Laos. Springer, Singapore, pp 99–116. https://doi.org/10.1007/978-981-96-2956-5_5

Phouxay K (2017) Patterns and consequences of undocumented migration from Lao PDR to Thailand. In: Bouté V, Pholsena V (eds) Changing lives in Laos: society, politics and culture in a post-socialist state. NUS Press, Singapore, pp 350–373

Rigg J (2013). From rural to urban: a geography of boundary crossing in Southeast Asia. TRaNS: Trans-Reg Natl Stud Southeast Asia 1:5–26. https://doi.org/10.1017/trn.2012.6

Rigg J, Salamanca A (2011) Connecting lives, living, and location: mobility and spatial signatures in Northeast Thailand, 1982–2009. Crit Asian Stud 43:551–575. https://doi.org/10.1080/14672715.2011.623522

Rigg J, Salamanca A, Parnwell M (2012) Joining the dots of agrarian change in Asia: a 25 year view from Thailand. World Dev 40:1469–1481. https://doi.org/10.1016/j.worlddev.2012.03.001

Rigg J, Salamanca A, Phongsiri M, Sripun M (2018) More farmers, less farming? Understanding the truncated agrarian transition in Thailand. World Dev 107:327–337

Sacklokham S, Chialue L, Yang F (2020) Rainfed and irrigated rice farming on the Savannakhet plain. In: Cramb R (ed) White gold: the commercialisation of rice farming in the lower Mekong basin. Palgrave Macmillan, Singapore, pp 151–168

Sato R, Jiang H, Nishimoto F, Yokoyama S (2025) Food procurement strategy of swidden agriculturalists in central Laos. In: Yokoyama S, Niwa T, Jiang H (eds), Population dynamics and livelihood changes of small-scale societies in Laos. Springer, Singapore, pp 19–43. https://doi.org/10.1007/978-981-96-2956-5_2

Scott JC (1976) The moral economy of the peasant: rebellion and subsistence in Southeast Asia. Yale University Press, New Haven

Shirai Y, Rambo AT (2017) Household structure and sources of income in a rice-growing village in northeast Thailand. Southeast Asian Stud 6:275–292. https://doi.org/10.20495/seas.6.2_275

Shirakawa C, Moriki Y (2025) Aspects of modern contraceptive use among women in rural northern Laos. In: Yokoyama S, Niwa T, Jiang H (eds) Population dynamics and livelihood changes of small-scale societies in Laos. Springer, Singapore, pp 207–222. https://doi.org/10.1007/978-981-96-2956-5_10

Shirakawa C, Nishimoto F, Takahashi S, Yokoyama S (2025) Growth in the use of modern contraceptive methods and the related social context in a central Laotian farming village. In: Yokoyama S, Niwa T, Jiang H (eds) Population dynamics and livelihood changes of small-scale societies in Laos. Springer, Singapore, pp 117–137. https://doi.org/10.1007/978-981-96-2956-5_6

Takahashi S, Yokoyama S, Nishimoto F, Niwa T (2025) Population growth and generational response to livelihoods in a rain-fed paddy farming village of central Laos. In: Yokoyama S, Niwa T, Jiang H (eds) Population dynamics and livelihood changes of small-scale societies in Laos. Springer, Singapore, pp 73–97. https://doi.org/10.1007/978-981-96-2956-5_4

Watanabe K, Hoshikawa K, Miyagawa S (2007) Extension of rainfed paddy field areas in Don Daeng village of northeast Thailand during 70 years. Nōgyo Nōson Kougakkai Ronbunshu

(Trans Jpn Soc Irrig, Drain Rural Eng) 251:507–513 (in Japanese). https://doi.org/10.11408/jsidre2007.2007.507

Yeung WJJ (2022) Demographic and family transition in Southeast Asia. Springer Nature, Cham, Switzerland. https://doi.org/10.1007/978-3-030-85679-3

Yokoyama S (2023) Special issue on population dynamics and livelihood changes of small-scale society in Laos. E-J GEO 18:258–263 (in Japanese). https://doi.org/10.4157/ejgeo.18.258

Yokoyama S, Takahashi S, Niwa T, Nishimoto F (2025) Population dynamics and paddy holdings in remote rural villages of northern Laos. In: Yokoyama S, Niwa T, Jiang H (eds) Population dynamics and livelihood changes of small-scale societies in Laos. Springer, Singapore, pp 141–170. https://doi.org/10.1007/978-981-96-2956-5_7

Livelihoods in a Swidden Agriculture Community

Food Procurement Strategy of Swidden Agriculturalists in Central Laos

Ren'ya Sato, Hongwei Jiang, Futoshi Nishimoto, and Satoshi Yokoyama

Abstract We conducted a study on the seasonal diet and food acquisition strategies of households in a hamlet inhabited by an ethnic minority (Mankhong) in Xepon district in central Laos, who maintain a subsistence economy by engaging in swidden agriculture, fishing, animal rearing, hunting, and gathering. We adopted the food diary method by which we collected the data of each day's side dish variety and regarding who acquired the main ingredients for the side dishes. It was concluded that the production-consumption balance in household food acquisition turns from negative to positive when the children are in their teens, suggesting that the contribution of children to household food acquisition as they grow up plays an important role.

Keywords Food diary · Food procurement · Household and life course · Subsistence activity · Laos

---

This chapter was a revised version of Sato et al. (2023).

R. Sato (✉)
Graduate School of Humanities, Osaka University, Toyonaka, Japan
e-mail: rsato7788@gmail.com

Present Address:
H. Jiang
National Institute for Environmental Studies, Tsukuba, Japan
e-mail: jianghw@gmail.com

F. Nishimoto
Embassy of Japan in Laos, Vientiane, Laos
e-mail: nsmtfts@gmail.com

S. Yokoyama
Graduate School of Environmental Studies, Nagoya University, Nagoya, Japan
e-mail: yokoyama.satoshi.i4@f.mail.nagoya-u.ac.jp

S. Yokoyama et al. (eds.), *Population Dynamics and Livelihood Changes of Small-Scale Societies in Laos*, International Perspectives in Geography 22,
https://doi.org/10.1007/978-981-96-2956-5_2

1 Introduction

This chapter's objective is to present an overview of the eating habits and food acquisition strategies of the Mankhong Mon-Khmer ethnic group of central Laos, who engage in swidden agriculture as their main subsistence, based on a year-long record of their meals and ingredient acquisition. Mankhong meals are centered on sticky rice, produced through swidden agriculture, with side dish ingredients (e.g., vegetables and proteins) obtained through fishing, animal husbandry, hunting, and gathering, as well as production of vegetables in swidden agriculture. Specifically, this study uses data on side dishes and those who acquired them, obtained through the food diary method described below, to clarify (1) the side dishes eaten throughout the year and quantities in which they were consumed and (2) who acquired the side dish ingredients, as well as from where and how they acquired it. Regarding the latter, the focus is on division of labor within the household and household life stages, in particular, considering the roles played in food acquisition by household members of different genders and ages, how these roles change in accordance with shifting life stages as household members grow up and age, and how the resulting energy flow among households at different life stages influences the survival strategies of the hamlet as a collection of households.

Classical research on the balance of household labor force and food consumption, as well as small-farm agricultural production based on the shifts therein, includes the work of the early 20th-century Russian agricultural economist Alexander V. Chayanov (Tomobe 1988). Chayanov focused on the fact that in nuclear family households children are born at fixed intervals, and with each child's growth the production–consumption balance (i.e., the value of the amount an individual produces minus the amount they consume) goes from negative to positive; he analyzed the changing phases of this balance between production and consumption along with household life stages. Needless to say, although children have a negative production-consumption balance while in early childhood, from a certain age they begin to gradually increase their contributions to the household's agricultural production, which maximizes just before they get married and become independent. Chayanov was the first to examine the household growth cycle in the analysis of small-farm household production capacities.

Chayanov's classical small-farm household studies have been followed by contemporary research on small-scale societies based on life history theory, which explains the timing of birth, growth, reproduction, and death of organisms through adaptation. The application of life history theory to humans has taken place since the 1990s, primarily among American and British biological anthropologists. Based mainly on the life history data of hunter-gatherers and swidden agriculturalists, these researchers have analyzed the correlations among growth and the timing of marriage, fertility and the timing of childbirth, age-related changes in subsistence contribution, and so on, revealing various characteristics of human life histories (Kaplan et al. 2000).

A characteristic of human life history research is its implication that life histories differ across cultures. For example, according to Bentley et al. (1993), hunter-gatherers and swidden agriculturalists tend to have lower fertility than settled farmers, a point explained by life history theory to the effect that their production–consumption balances differ. For instance, hunter-gatherers, who depend on hunting large animals, require a long time to acquire hunting skills, so the time until adulthood is longer, with marriage and childbirth likewise occurring later. Higher ages of marriage and first childbirth have a significant effect on total fertility rates, meaning that hunter-gatherers tend to have lower fertility levels. As for swidden agriculturalists, although they are farmers in the sense that they depend on swidden crops for their subsistence, many of them spend a relatively high proportion of their working time on hunting and gathering because swidden cropping is hard to master at a young age. They are similar to hunter-gatherers in the timing of their acquisition of subsistence skills and the shifts in production-consumption balance (Sato 2014).

However, caution should be exercised when speaking of hunter-gatherers or swidden agriculturalists as a single group. The former acquire food in diverse ways appropriate to their environments, from using spears or guns to hunt large animals to trapping smaller ones, involving very different levels of skill; there are also significant differences between groups relying mainly on hunting or fishing and those who are mainly gatherers. Swidden agriculturalists likewise vary widely in their occupations outside swidden agriculture. Therefore, discussions must take these variations into account. For example, Sellen and Mace (1997) focus on the degree of dependence on agriculture within each group and find a positive correlation between the degree of dependence on agriculture and fertility.

In this context, Kramer and her colleagues discuss the relations among household earning strategies, children's growth, and fertility based on detailed subsistence data from the Maya swidden agriculturalists, who depend highly on agriculture (Kramer 2005a, b; Kramer and Ellison 2010; Kramer and McMillan 2006). According to them, among Maya people, who generally acquire skills at a relatively early age, children acquire numerous skills in stages, including caring for younger siblings. As a result, their production-consumption balance shifts from negative to positive at an early age, and they contribute significantly to the household budget until they marry and leave the household. Kramer and her colleagues argue that this explains the short birth spacing and, ultimately, the high total fertility rate of the Maya people.

This discussion of individual and household production-consumption balance requires consideration of the flow of food production, which is not complete within a single household but takes place between different households and generations as well. For example, according to research on the Aché hunter-gatherers of Paraguay, most of the meat from large animals, which is their main source of calorie, is obtained by adult males. However, hunting techniques take time to develop, with production peaking for men in their 30 and 40s. For Aché men, the household production–consumption balance finally reaches a positive value in their 20s, and in their 30 and 40s, at the peak of production, they provide food not only for their own children but also for others outside the households, as well (Kaplan et al. 2000).

In addition, prior to their life history research, Kaplan and Hill (1985) reported case studies regarding how the products of hunting and gathering were shared among the group. A survey on who acquired each ingredient and who ate it for each category of hunted and gathered food among the Aché revealed that almost all the large-animal meat was eaten by group members other than the hunter who had acquired it, and most of the gathered nuts, shellfish, and so on were consumed within the household. Although predictable, these results show that rare acquisitions such as large-animal meat are actively shared among the group, whereas the results of proportional effort such as gathering tend to be limited to exchanges within households or among relatives. These previous studies suggest that division of labor in occupations within households and the flow of food between households are constrained by the level of acquisition skills required for each type of food.

Results from a national nutrition survey conducted in Laos indicated that adult men and women generally have insufficient intakes of all types of nutrients, except sodium (Ratsavong et al. 2020). Food insecurity remains a serious problem. Understanding the balance between food production and consumption at the household level, the minimal unit of food production and consumption, is essential to understanding why these macroeconomic outcomes have occurred. Previous research in Laos has attempted to shed light on these issues through the analysis of long-term food diaries in food security studies in rural Laos (Hasada and Yamada 2017; Nonaka et al. 2014). In particular, Nonaka's team analyzed the frequency of occurrence of sources (wild or cultivated) and categories (animal or plant) of side dishes in three villages in northern, central, and southern Laos with different locations and livelihoods, collecting food diaries from 20 households in each village for one year. However, they did not include in the analysis who produced (collected) and who consumed the food. In other words, there was insufficient analysis of the division of labor within households and the flow of food among households, an important strategy for ensuring household food security in terms of food acquisition and redistribution.

Based on the above discussion, this chapter provides an overview of the Mankhong people's meals over a year, focusing on side dish ingredients other than sticky rice, which is the main source of sustenance. The study categorizes the main side dish ingredients, clarifies who (inside or outside the household) acquires the food eaten in each household, and discusses the Mankhong's food acquisition and earning strategies. Sticky rice was excluded from the survey because, as noted below, it was impossible to obtain highly reliable data for it; moreover, the sticky rice stored in each household was considered unlikely to be a part of inter-household exchanges. The following section provides an overview of the study area, and Sect. 3 discusses data collection and analysis methods. Section 4 reports on the side dish varieties in the study hamlet and their seasonal changes, and Sect. 5 analyzes the acquirers of side dish ingredients by household stage. Based on these results, Sect. 6 reports on food acquisition strategies among the Mankhong, and Sect. 7 provides a summary of the results.

2 Study Area

This survey targeted AN hamlet[1] in Xepon District, Savannakhet province, in central Laos (Figs. 1 and 2). The inhabitants of the hamlet are Mankhong, a minority group who speak a Mon-Khmer language; as of February 2020, there were 29 households and 146 people. Although the hamlet is located on the main road from northeastern Thailand across the Mekong River, connecting the central Lao hub of Savannakhet to Hue in Vietnam via Xepon, the main road area and AN hamlet are separated by the Xepon River, with no connecting bridge. Therefore, the residents travel back and forth by boat (Fig. 3). Consequently, the development of the market economy has been relatively slow compared to the roadside hamlet; as of 2020, almost all the ingredients in AN hamlet meals were procured by AN residents. AN hamlet is located straight up the steep southern bank of the Xepon, and river water is used for drinking and everyday use (until February 2020, when the survey took place; thereafter, a pump-type well was dug in the hamlet, with drinking water, etc., being extracted from the well as of September 2022). Across the river is the elementary school; while AN hamlet children do attend school, observation suggests that the enrollment rate is not high. Electrification has also been slow, with some households relying on car batteries, etc., to power household appliances such as televisions until electricity first arrived in 2019.

The main subsistence in AN hamlet, as mentioned above, is swidden agriculture. Most of the crops are a sticky variety of dry rice, which is the main staple food of the residents. Vegetables (e.g., eggplant, chili peppers, squash, beans, taro, leafy vegetables) are grown at the edges of rice fields and used as side dish ingredients. The rice fields are slashed in February and March and planted after drying and burning. The adult men and women of the hamlet do the slashing work, and weeding is conducted at least three times before harvest, mainly by women. The rice is harvested in October, by which time the vegetable harvest is also complete. The chaff is removed by hand, and the rice is placed in a basket attached to the waist for harvesting. The swidden normally lies fallow after a year of cropping and is slashed again after six years or more. In many of the fallow forests, bamboo grows abundantly as a pioneer species.

The wide variety of domestic animals includes cattles, water buffalos, goats, pigs, chickens, ducks, dogs, and cats. Animals other than dogs and cats are raised for sale and also eaten at hamlet events and festivals. Some households also eat dogs. Most households kept a small number of these animals, but only one household was engaged in cattle fattening (pastoralism) as its main subsistence.

Fishing takes place in the Xepon River, just below the hamlet, and the aquatic resources derived from the river are the main daily source of protein. In the river, freshwater fish are caught with whiskers and fishing rods, and shellfish, shrimp and crabs are gathered in the shallows. Fishing with whiskers and fishing rods is mainly

[1] AN and AG hamlets, on the opposite bank of the Xepon River, form an administrative division as a village. Because this study did not target AG hamlet for study, the term "hamlet" rather than "village" is used.

Fig. 1 Location of AN hamlet

a male activity, although boys in their teens are also actively involved. On the other hand, shellfish and crustacean gathering is often done by women, and girls under the age of 10 also participate in shellfish gathering, especially during the dry season.

The swiddens and surrounding fallow forests are sources of bamboo shoots, mushrooms, and wild herbs, as well as wild birds, rodents such as squirrels and mice, reptiles such as lizards, and mammals such as bats and flying squirrels, all of which are trapped for eating. Frogs and insects such as crickets are also gathered and used as ingredients in main dishes. Hunting is mainly done by men and gathering of bamboo

Fig. 2 AN hamlet

Fig. 3 A ferry crossing the Xepon River

shoots, mushrooms, and wild herbs is mainly done by women. As with the work at the river, the work of hunting-gathering tends to be divided along gender lines. Furthermore, women perform everyday water carrying, rice hulling and refining with a mortar and pestle, and cooking.

Although cash is obtained through selling the gathered bamboo shoots and leaving the hamlet to work on farms, etc., most food ingredients are supplied within the hamlet. However, Salt and flavorings are purchased, meat is sometimes bought and sold within the hamlet, and sticky rice is also purchased when the swidden harvest is insufficient.

3 Research Materials and Methods

Dietary and ingredients acquirer data were collected using the food diary method (Nonaka et al. 2014). Fourteen informants of different ages and genders were selected from 14 different households and asked to keep a daily record of each meal. Recording in the diary sheets was entrusted to two survey assistants living in the hamlet, who visited each household daily to record their meals. Using this method, we were able to obtain records of 13,934 meals between May 2015 and November 2017, which were used for the analysis in this study.

The diary format used one line for each meal record, including the date and time of day (chosen from numbers indicating five possible periods: morning, mid-morning, noon, afternoon, and night). The design included with whom the meal was eaten, where it was eaten, the approximate amount of steamed rice eaten, the full menu of side dishes, who acquired the main ingredients of the side dishes (acquirer), and where the ingredients were acquired (selected from numbers indicating several possible options: within the hamlet, the fields, the woods, the river, and so on).

Almost all meals eaten in AN hamlet consist of steamed rice, a main side dish, and one or two other side dishes (Fig. 4). In some cases, the meal consisted only of rice and the main dish, and in other cases, there are additional side dishes such as a chili pepper dip for rice (jeow, made by crushing raw peppers and mixing them with salt, lime, etc.), boiled bamboo shoots, raw eggplant, and so on. It is also not uncommon for meals to lack a main side dish and consist only of rice and additional side dishes. As the menus below show, most side dishes are made from a single main ingredient, which makes it easy to identify the main ingredients in each menu.

The term "acquirer" here refers to the person who acquires the food by harvesting, fishing, hunting, gathering, and so on. When crops were gifted or purchased within the hamlet, the gifter or seller was indicated in the record. Thus, by recording the details of the acquirer of the ingredients of meals provided within the household, it became possible to clarify the distinction between the household and the outside household: who acquired each ingredient and who ate it.

For traditional Lao meals, the custom is that steamed sticky rice is brought to the table in a small basket; each person scoops up a handful of rice, rolls it into a ping-pong ball sized shape, and adds side dishes as they eat. To estimate how much

Fig. 4 A sample meal of steamed rice, mushroom salad, and chili pepper dip

steamed rice individuals were eating, informants were asked to record roughly how many handfuls of rice they took. However, almost the same number was recorded for each meal, suggesting that the data were unreliable. To supplement these figures, the steamed rice in the basket on the table was weighed before and after meals during the on-site survey to calculate the steamed rice intake. However, due to the insufficiency of the data, this analysis focuses only on the side dish menus without specifying the steamed rice intake.

In addition to the diary records, in order to obtain quantitative information on the actual ingredients, one of the informants assigned to the recording process was provided with a digital camera and asked to photograph each meal as much as possible (Fig. 5). Although constant photographic data could not be obtained, the photographs provided significant information about the ingredients.

Although the meal records targeted individuals, it is assumed that, except for gatherings or ceremonies where invited guests ate together, the entire household generally consumed the same food prepared by the adult women of the household. Thus, the individual records were used in the analysis as records of the entire household's meals. For this reason, care was taken in the selection of informants (diarists) to include households in different stages. Of the 14 informants selected, three provided only short-term records because they married out during the study. Therefore, although the meal menu aggregation and analysis included the records of 14 informants, the household analysis used data from 11 households, excluding these 3.

The 11 households included in the analysis were composed as shown in Table 1 (age as of 2016 for reference). Since the analysis in this study focuses on the presence and age of children, the age of children in the households were used as reference for classification into three groups: (1) households of young couples without children or with children under 10 years old; (2) households with children aged 10–19 years; and (3) older couples with children living independently. The basic hamlet household format consists of nuclear families centered on married couples. The Mankhong are patrilineal and tend to be virilocal, so many of the women had married from outside the hamlet, and their husbands' brothers and parents often lived in the same hamlet.

4 Side Dish Menus and Seasonality

The main source of calories for AN residents is steamed rice; their eating habits can be characterized as "large quantities of sticky rice with frugal side dishes." They generally have meals in the morning, noon, and evening, but steamed rice is cooked once a day and consumed by the family throughout the day. Side dishes[2] serve as accompaniments to the steamed rice and provide protein and other nutrients.

---

[2] Instances of multiple side dishes have been divided into "main," considered the most important, and others. When meals consisted only of steamed rice and one side dish, that dish was labeled "main."

a. Raw eggplant and chili pepper dip

b. Frog soup

c. Roasted mouse

d. Boiled bamboo shoots and chili pepper dip

e. Bamboo shoots soup

f. Clam soup

g. Mushroom soup and chili pepper dip

h. Freshwater fish soup

Fig. 5 Examples of menus photographed by informants

Table 1 Family composition of households surveyed

Household no.	Start date of diary	End date of diary	No. of records	Household type	Remarks
The households with children under 5 years of age					
2	2015/3/24	2017/8/30	1,479	20–30_mf 5_m·5_f	5_f was born in Jun, 2016
3	2015/5/18	2017/8/29	1,635	60_f 30_mf 5_m·5_f	
8	2015/5/15	2017/9/4	879	20_mf 5_m·5_m·5_f	
10	2016/6/5	2017/9/4	881	10–20_mf 5_f	A second daughter was born in Jun, 2017
11	2016/6/5	2017/3/24	334	40_mf 10–20_mf 5_f	5_f was born in Jan, 2017
The households with teenagers					
1	2015/4/27	2017/8/29	1,667	60_f 40_mf 10M_m·10L_f·10L_m	
4	2015/5/17	2017/8/27	1,711	30_mf 10M_m·10L_m·10L_m	
5	2015/5/18	2017/8/31	1,662	40_mf 10H_f·10M_m	
7	2015/5/16	2017/9/4	979	40–50_mf 10H_m 10H_f	10H_m was married in Feb, 2017 and his wife was joined
9	2016/5/27	2017/8/30	798	40–50_mf 10H_f·10L_f·10L_f	
The household with independent children					
6	2015/5/16	2017/9/4	931	50_mf	

Source Authors' interview survey
Note Symbols and figures of family composition are as follows. 30 refers to ages in the 30s, 5 refers to a 5 year-old child. m is for male and f for female. For teenagers, L, M, and H are used to mean early, middle, and late teens, respectively

Table 2 shows the frequency of (main) side dishes served for each cooking process. Soups account for approximately 42% of the total. Other cooking methods include

Table 2 Frequency of appearance by dish type

Dish category	Frequency	%
Soups	5,815	41.7
Boiled foods	1,693	12.2
Mixed foods	1,547	11.1
Roasted foods	1,260	9.1
Larb	487	3.5
Fermented vegetables, fish, or meat	381	2.7
Jeow	359	2.6
Dried meat or fish	143	1.0
Steamed foods	46	0.3
Stews	2	0.0
Other	2,201	15.8
Total	13,934	100.0

Source Created based on food diary data

boiling, mixing, roasting, and larb[3] amounting to 36% in total; some percentage consists of simple side dishes of fermented vegetables, meat, or fish, as well as chili pepper dip alone. Typical meals in AN hamlet consist of rice, soup, and side dish, sometimes with two or three side dishes. Foods prepared with cooking oil are rare, the side dishes are generally low in fat and calories.

Frequent menus based on cooking methods are shown in Table 3. Fish, frogs, and meat are often simply roasted, while vegetables and bamboo shoots are simply boiled. Except for boiled foods, fish is used as an ingredient extremely often. Similarly, except for roasted foods, bamboo shoots are used frequently.

Fish throughout the year and bamboo shoots during the rainy season may be considered the typical side dishes of AN hamlet.

Let us look at the frequency of main side dishes by ingredient (Tables 4 and 5). Main ingredients are evenly divided between animal and vegetable (including mushrooms) ingredients (Table 4). Within these broader categories, approximately 80% consists of fish, crops (vegetables), bamboo shoots, and meat of domestic and wild animals (e.g., frogs, lizards, and mammals such as mice and squirrels; Table 5). In particular, fish, field crops, and bamboo shoots account for just under 60% of the total, confirming that fish and bamboo shoots are the most important ingredients. Other important ingredients include shellfish, shrimp, and crabs gathered from the river, as well as mushrooms and other gathered ingredients; insects and eggs are rarely consumed. In contrast to Lao people, who like to eat insects, Mankhong people are not especially inclined to do so.

[3] Larb is a Lao side dish in which meat or fish is finely chopped, lightly heated, and mixed with herbs, lime, etc. It was categorized separately rather than as a mixed or roasted dish although both categories could apply.

Table 3 Frequent menus by type of dish

Soups	**Frequency**
Fish soup	1,273
Bamboo shoot soup	1,210
Snail soup	540
Mushroom soup	537
Chicken soup	368
Frog soup	307
Bong (aquatic plant) soup	274
Rattan soup	189
Small frog soup	171
Wild mouse soup	147
Boiled foods	**Frequency**
Boiled vegetables	648
Boiled bamboo shoots	367
Boiled banana flowers	199
Boiled greens	144
Boiled bananas	129
Mixed foods	**Frequency**
Smashed and mixed bamboo shoots	256
Fish mixed with herbs	200
Wild mouse mixed with herbs	173
Pork mixed with herbs	149
Smashed and mixed eggplants	114
Roasted foods	**Frequency**
Grilled fish	579
Steamed and roasted fish	132
Grilled chicken	70
Grilled pork	78
Grilled small frog	81

Source Created based on food diary data

Table 4 Frequency of appearance of animal and vegetable menus

Ingredient category	Frequency	%
Animal	7,110	51.0
Vegetable and mushroom	6,642	47.7
Other	182	1.3
Total	13,934	100.0

Source Created based on food diary data

Table 5 Frequency of appearance by ingredient category

Ingredient category	Frequency	%
Fish	3,164	22.7
Crops (other than rice)	2,936	21.1
Bamboo shoots	2,116	15.2
Meat	1,531	11.0
Wild herbs	1,041	7.5
Amphibians and reptiles	831	6.0
Wild mammals	774	5.5
Shellfish and crustaceans	679	4.9
Mushrooms	549	3.9
Other	182	1.2
Insects	91	0.7
Eggs	40	0.3
Total	13,934	100.0

Source Created based on food diary data

The frequency (percentage) of each category by month is shown in Fig. 6.[4] Seasonality is comparatively clear for bamboo shoots, mushrooms, frogs, and lizards, which are eaten mainly in the rainy season; shellfish and wild herbs, which are eaten in the dry season; and insects such as crickets, which are eaten particularly in September and October. Meanwhile, commonly eaten fish and field crops tend to be consumed throughout the year. Meat and eggs are consumed more frequently in August, September, and the dry season. However, since meat is often served at ceremonies, this may be related to the timing of ceremonies. The frequent appearance of fish on the menu is explained by its availability throughout the year.

5 Strategies for the Acquisition of Side Dish Ingredients

This section presents the strategies used by households to acquire side dish ingredients. The acquisition strategies of the main primary ingredients for main side dishes are given in Table 6, divided by husband (householder), wife, child(ren), husband's parents, relatives outside the household (e.g., husband's siblings, etc.), and non-relatives[5] (e.g., neighbors).

Let us confirm the main roles by category of acquirer. The husband's primary role is fishing, which produces over half of the total food acquired. This is followed by

[4] Field crops are more frequent in November, specifically including potatoes, green beans, eggplant, squash, and so on. However, the records for November were scarce compared to other months (179 out of 13,934), so sample error must be considered.

[5] Acquirers listed in the diaries as "residents" or "friends" were considered to be non-relatives.

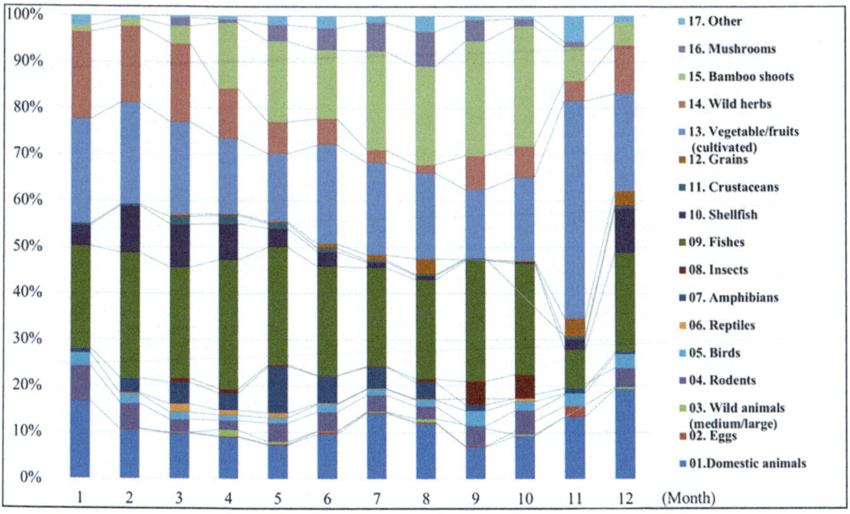

Fig. 6 Seasonal variation by ingredient. *Source* Created based on food diary data

field harvesting and hunting of wild mammals, mainly by trapping. The husband is also responsible for catching frogs and wild birds. Meanwhile, the wife's primary roles include field harvesting and bamboo shoot gathering, followed by gathering of wild herbs, shellfish, crustaceans, and mushrooms, fishing, and catching frogs. This indicates that adult men are mainly involved in fishing and hunting and adult women are mainly involved in field harvesting and gathering at the river and fallow forests.

Next, in terms of children's[6] work, the main roles are fishing and picking bamboo shoots, which are assumed to be the work of boys and girls, respectively. These are followed by other important roles such as field harvesting, catching frogs, and gathering shellfish and crustaceans. Based on the adults' roles, it is assumed that girls are mainly responsible for these tasks. Other roles include gathering wild herbs and mushrooms, catching insects,[7] and catching mammals. Although hunting was assumed to be boys' work, interviews revealed that children's contribution to trapping and hunting was minimal, with full-scale hunting being done by adult men. Overall, boys play a major role in supplying protein to the household through fishing and girls by gathering shellfish and catching frogs. Further, based on observation and interviews, fishing and catching frogs are done by older children and gathering shellfish by younger children. This suggests that girls tend to start contributing to food acquisition earlier than boys.

[6] Because the diaries frequently make it difficult to identify individuals among siblings, no gender division has been made.

[7] As the data shows, eating insects is uncommon in AN hamlet. The households where insects were frequently eaten included that of a Lao wife who had married in. Her children (sons) were observed to be gathering crickets, etc.

Table 6 Frequency of appearance by ingredient category viewed by acquirer category

Ingredients acquired by householder (husband)	**Frequency**	**%**
Wild mammals	398	13.1
Wild birds	169	5.5
Reptiles	36	1.2
Amphibians	242	7.9
Insects	8	0.3
Fish	1,560	51.2
Shellfish and crustaceans	26	0.8
Field crops	414	13.5
Wild herbs	24	0.8
Bamboo shoots	55	1.8
Mushrooms	39	1.3
Other	78	2.6
Total	3,049	100.0
Ingredients acquired by householder (wife)	**Frequency**	**%**
Wild mammals	17	0.4
Wild birds	3	0.1
Reptiles	20	0.4
Amphibians	109	2.5
Insects	33	0.8
Fish	268	5.9
Shellfish and crustaceans	336	7.7
Field crops	1,359	31.1
Wild herbs	662	15.1
Bamboo shoots	1,204	27.5
Mushrooms	317	7.3
Other	53	1.2
Total	4,371	100.0
Ingredients acquired by children	**Frequency**	**%**
Wild mammals	30	1.6
Wild birds	10	0.5
Reptiles	4	0.2
Amphibians	197	10.7
Insects	33	1.8
Fish	514	27.8
Shellfish and crustaceans	194	10.5
Field crops	263	14.2
Wild herbs	79	4.3

(continued)

Table 6 (continued)

Ingredients acquired by children	**Frequency**	**%**
Bamboo shoots	460	24.9
Mushrooms	55	3.0
Other	10	0.5
Total	1,849	100.0
Ingredients acquired by householder's parents	**Frequency**	**%**
Purchased meat, eggs, etc.	97	9.1
Wild mammals	8	0.7
Wild birds	1	0.1
Reptiles	1	0.1
Amphibians	16	1.5
Insects	14	1.3
Fish	179	16.7
Shellfish and crustaceans	72	6.7
Field crops	221	20.7
Wild herbs	165	15.4
Bamboo shoots	208	19.5
Mushrooms	80	7.5
Other	7	0.7
Total	1,069	100.0
Ingredients acquired by relatives not living in the household	**Frequency**	**%**
Purchased meat, eggs, etc.	401	42.6
Wild mammals	36	3.8
Wild birds	4	0.4
Reptiles	8	0.8
Amphibians	11	1.2
Fish	132	14.0
Shellfish and crustaceans	20	2.1
Field crops	144	15.3
Wild herbs	71	7.5
Bamboo shoots	61	6.5
Mushrooms	13	1.4
Other	41	4.4
Total	942	100.0
Ingredients acquired by non-relatives	**Frequency**	**%**
Purchased meat, eggs, etc.	198	17.6
Wild mammals	36	3.2
Wild birds	7	0.6

(continued)

Table 6 (continued)

Ingredients acquired by non-relatives	**Frequency**	**%**
Reptiles	4	0.3
Amphibians	51	4.5
Insects	1	0.1
Fish	283	25.2
Shellfish and crustaceans	20	1.8
Field crops	391	34.8
Wild herbs	16	1.4
Bamboo shoots	57	5.1
Mushrooms	5	0.4
Other	56	5.0
Total	1,125	100.0

Source Created based on food diary data

Next, foods acquired by the husband's parents include, in the order of frequency, field crops, bamboo shoots, fish, and wild herbs. All of these are frequently served as side dishes.

In the case of young couples without children or with children under 10 years of age, the generation of the householder's parents tends to act as acquirer. In this case, the parents themselves are still young and fully capable of contributing to food acquisition, so it can be interpreted as directing this surplus to their sons' households with young children. Given that parents also frequently provide their sons' households with purchased meat or eggs, we estimate that the parental generation is relatively comfor l at this stage.

Next, let us look at relatives (excluding parents) outside the household, that is siblings, aunts, uncles, and so on, who reside in the hamlet. Notable here is the very high frequency of meat or eggs sharing. This suggests that when animals are butchered in the hamlet, relatives share the meat or purchase it from one another. Field crops and fish are also represented, but a lower rate than parents' provisions, suggesting some exchange between relatives other than parents.

Finally, we have foods provided by non-relatives. These also include field crops and fish, notably at a rate not much lower than those provided by relatives. The provision (purchase) of meat and eggs is also frequent, although not as frequent as between relatives. When comparing the exchange of food ingredients between relatives and between non-relatives, we find that while the former is heavily weighted toward the provision and purchase of meat and eggs, the latter plays just as important a role via the gifting of vegetables and fish, which are consumed on a daily basis.

To confirm the above conjectures on roles within and outside the household, Tables 7, 8 and 9 show an aggregation of acquirer data by household type for the informants. First, in households with no children or with children under the age of 10, the frequency of acquisition by the husband and wife was extremely high, ranging from

44.2 to 75.6%. In particular, households 8 and 10, in which the husband and wife's acquisition rates were high, received almost no food from parents or other relatives. Instead, they had a relatively high rate of receiving food from non-relatives at 15–19%. Meanwhile, households 2 and 11 received large amounts of food from parents and other relatives, and none at all from non-relatives. Household 3 was somewhere in the middle, receiving some amount from both relatives and non-relatives. Thus, households where children cannot contribute to household subsistence or where care of young children is required receive some amount of food from parents or from neighbors otherwise.

Table 7 Frequency of appearance by acquirers in young couples' households

Acquirer	Frequency	%
Households 2		
Householder (wife)	564	38.2
Householder (husband)	427	28.9
Gift (relatives)	415	28.1
Purchase/household livestock, etc.	58	3.9
Child(ren)	14	0.9
Total	1,478	100.0
Households 3		
Householder's mother	462	24.3
Householder (husband)	435	22.9
Householder (wife)	405	21.3
Gift (relatives)	242	12.7
Gift (non-relatives)	193	10.2
Purchase/household livestock, etc.	145	7.6
Unclear	18	1.0
Total	1,900	100.0
Households 8		
Householder (wife)	346	39.4
Householder (husband)	318	36.2
Gift (non-relatives)	137	15.6
Purchase/household livestock, etc.	45	5.1
Gift (relatives)	21	2.4
Unclear	12	1.3
Total	879	100.0
Households 10		
Householder (wife)	422	47.9
Householder (husband)	237	26.9
Gift (non-relatives)	168	19.0

(continued)

Table 7 (continued)

Acquirer	Frequency	%
Purchase/household livestock, etc.	44	5.0
Gift (relatives)	5	0.6
Unclear	5	0.6
Total	881	100.0
Households 11		
Householder (wife)	182	54.5
Husband's father	66	19.7
Householder (husband)	39	11.7
Husband's mother	30	9.0
Unclear	8	2.4
Gift (relatives)	5	1.5
Child(ren)	2	0.6
Purchase/household livestock, etc.	2	0.6
Total	334	100.0

Source Created based on food diary data

Table 8 Frequency of appearance by acquirer in households with children aged 10–19 years

Acquirer	Frequency	%
Households 1		
Child(ren)	611	36.7
Householder (wife)	472	28.3
Householder (husband)	419	25.1
Gift (relatives)	92	5.5
Purchase/household livestock, etc.	57	3.4
Gift (non-relatives)	10	0.6
Unclear	6	0.4
Total	1,667	100.0
Households 4		
Householder (wife)	427	25.0
Gift (relatives)	388	22.7
Child(ren)	317	18.5
Householder (husband)	243	14.2
Purchase/household livestock, etc.	169	9.9
Gift (non-relatives)	165	9.6
Unclear	2	0.1
Total	1,711	100.0

(continued)

Table 8 (continued)

Acquirer	Frequency	%
Households 5		
Householder (husband)	588	35.4
Householder (wife)	497	29.9
Child(ren)	435	26.2
Purchase/household livestock, etc.	74	4.4
Gift (relatives)	68	4.1
Total	1,662	100.0
Households 7		
Householder (husband)	217	22.2
Householder (wife)	202	20.7
Child(ren)	153	15.6
Gift (non-relatives)	151	15.4
Son's wife	150	15.3
Purchase/household livestock, etc.	97	9.9
Unclear	7	0.7
Gift (relatives)	2	0.2
Total	979	100.0
Households 9		
Householder (wife)	335	39.5
Child(ren)	237	28.0
Householder (husband)	110	13.0
Gift (relatives)	107	12.6
Purchase/household livestock, etc.	49	5.8
Gift (non-relatives)	9	1.1
Total	847	100.0

Source Created based on food diary data

Next, in households with children aged 10–19 years, the children's food acquisition rate was 15.6–36.7%, playing an important role in the household budget. The difference in the role of children may be due to the age and number of children over 10 years. Household 4, where the householder couple was the youngest, retained some of the characteristics of households headed by young couples, with the children's acquisition rate on the low side at 18.6% and provision from relatives at 22.7%. Conversely, household 1, which had the highest child acquisition rate at 36.7%, relied on the couple and their children for more than 90% of its ingredients, with minimal reliance on relatives or neighbors. Households at this stage are likely to be the most productive, with surplus ingredients likely to be given to relatives and

Table 9 Frequency of appearance by acquirer in households with independent children

Households 6		
Acquirer	Frequency	%
Householder (wife)	413	44.4
Householder (husband)	278	29.9
Gift (non-relatives)	157	16.8
Purchase/household livestock, etc.	50	5.4
Gift (relatives)	29	3.1
Unclear	4	0.4
Total	931	100.0

Source Created based on food diary data

other neighbors. Households 5, 7, and 9 were in the middle, receiving food from relatives or non-relatives at a low rate. In these households, the first one or two children had already become independent and food was provided to young couples. Overall, this category of households is likely to be positioned at the stage where household production-consumption balance (see Sect. 1) is shifting from negative to positive in accordance with the growth of children.

Finally, let us consider older couples whose children are independent (household 6). At this stage, the contribution rate of husband and wife increases again to 74.3%. Other contributions include those from non-relatives at 16.9% and relatives at 3.1%. The latter is assumed to include contributions from son households.

Couples in their 50s are assumed to have sons in their 30s or younger, with the oldest child's household at the turning point to positive production-consumption balance and those of younger children still at the negative stage. Although couples in their 50s are assumed to have returned to negative production-consumption balance, their children's households are not yet ready to compensate for this, suggesting that, they depend to a small extent on provision of non-relatives, such as neighbors.

6 Discussion

Changes in food acquisition strategies in accordance with household stages were clarified through aggregation of ingredient acquirer data from the food diaries. The key to explaining these changes is the varying degree of contribution to household subsistence based on children's age. As children grow older, they begin by helping with relatively low-skilled subsistence work, such as shellfish gathering, and gradually acquire the skills for bamboo and wild herb gathering, frog hunting, fishing with rods and whiskers, trapping, and hunting. Young couples whose children are too young to contribute, and elderly couples whose children are independent have to acquire almost all the side dish ingredients on their own; households with young children are particularly in need of resources, so support from parents and neighbors is thought to be important.

Meanwhile, households in which all children are aged between 10 and 19 years are at the most productive stage with resources to spare. At this stage, the couples at the center of the households are in their 30s or 40s, at the peak of their productivity. These households are expected to provide food for other relatives and neighbors in the village. Households with young children have parents, aunts, uncles, and neighbors who compensate for their negative production-consumption balance, and they return the favor or pay it forward to the next generation when their children begin to provide a surplus after the age of 10.

The presence of a flow of ingredients between unrelated households is noteworthy. Among the ingredients exchanged, the frequency of field crops (vegetables) is high, suggesting that "ingredients acquired through effort" are given as gifts between unrelated households at different stages of reciprocal relationships. Since the households receive larger quantities from non-relatives are in situations where relatives cannot provide the same, these relationships seem to function as a safety net.

The changing role of children in the household economy helps to explain the life histories of the AN residents, particularly the relatively early age of marriage for both men and women. For example, although the Majangir people of Ethiopia are also swidden agriculturalists, they have numerous advanced traditional techniques needed for household maintenance, such as gathering honey with honey traps in trees and hunting large animals, as well as pottery made by women. These techniques are reflected in a higher age at first marriage, resulting in a lower total fertility rate (Sato 2009, 2014). In contrast, the only AN hamlet subsistence that require advanced skills are swidden slashing and hunting; most subsistence can be mastered to some extent by children aged 10–19, thus rendering earlier marriage effective. Even in so-called swidden agricultural socities, which were charactierised by a combination of subsistence activities such as hunting and gathering, the comparision of two groups suggests that variations in the subsistence combinations between groups lead to variations in fertility.

7 Conclusion

Based on food diaries from 11 households in AN hamlet, this study has clarified the contents of their meals throughout a year, taking into account seasonal variations, and focused on the acquirers of major ingredients to discuss changes in food acquisition strategies according to household life stages. In terms of household food acquisition strategies, the relationship between household life stages and production-consumption balance has been clarified to some extent, revealing that young couples compensate for their negative production-consumption balance by providing food from parents and other relatives, that production-consumption balance shifts to positive when children are between 10 and 19 years old and surplus food is provided to the households of independent children, parents, or neighbors, and that the production-consumption balance returns to negative when children become fully independent, with households gradually relying on provision from the households of children or neighbors.

Due to insufficient data, this chapter could not include the production and consumption of the main staple food, rice, in the analysis. To analyze meals from the perspective of nutrient intake, including the amount of rice consumed per day, a survey of rice consumption is essential. This is a task for future research.

In addition, data on past fertility in AN hamlet show relatively low values, although both men and women married early. As discussed in the review at the beginning of this study, life history theory suggests that a high degree of contribution from children, early marriage, and high fertility are all interrelated. Thus, the analysis results in this study are not consistent with the low fertility rate of AN hamlet. These factors require a comprehensive analysis of the life history analysis of the residents, not limited to their food acquisition. This is another issue that need to be addressed in the future.

Acknowledgements This study was supported from JSPS Grants-in-Aid JP25257004 and JP17H01633. The authors are grateful to the cooperation provided in the on-site survey by the Lao Ministry of Health, Savannakhet Provincial Department of Health, and Savannakhet Province Xepon District Health Office. The authors would like to thank Dr. Sengchanh Kounnavong, Dr. Somphou Sayasone, and Dr. Tiengkham Pongvongsa for their advice to our research plan, The authors would also like to thank Dr. Inthanomchanh Vongphoumy, Dr. Phet Nansounthavong, Mr. Monethong Inthisth, Mr. Lamsee Phanthakeo, and Mr. Hongsa Sinpaseuth for their good help during the field survey period. Further, the authors express their deepest gratitude to the AN residents who kept detailed daily food diaries containing vast quantities of information. Their participation is sincerely appreciated.

References

Bentley G, Goldberg A, Jasienska G (1993) The fertility of agricultural and non-agricultural traditional societies. Popul Stud 47:269–281. https://doi.org/10.1080/0032472031000147006

Hasada K, Yamada R (2017) Raosu chubu ni okeru shokuryou nyusyu no genjou to kadai (Current status and issue on food access in a semi-mountainous village of central Laos). Water, Land Environ Eng (Suido no chi) 85(5):469–474 (in Japanese). https://doi.org/10.11408/jjsidre.85.5_469

Kaplan H, Hill K (1985) Food sharing among Ache foragers: tests of explanatory hypotheses. Curr Anthropol 26:223–246. https://doi.org/10.1086/203251

Kaplan H, Hill K, Lancaster J, Hurtado AM (2000) A theory of human life history evolution: diet, intelligence and longevity. Evol Anthropol 9:156–185. https://doi.org/10.1002/1520-6505(2000)9:4%3C156::AID-EVAN5%3E3.0.CO;2-7

Kramer KL (2005a) Maya children: helpers at the farm. Harvard University Press, Cambridge

Kramer KL (2005b) Children's help and the pace of reproduction: cooperative breeding in humans. Evol Anthropol 14:224–237. https://doi.org/10.1002/evan.20082

Kramer KL, Ellison PT (2010) Pooled energy budgets: resituating human energy allocation trade-offs. Evol Anthropol 19:136–147. https://doi.org/10.1002/evan.20265

Kramer KL, McMillan GP (2006) The effect of labor-saving technology on longitudinal fertility changes. Curr Anthropol 47:165–172. https://doi.org/10.1086/499550

Nonaka K, Ono E, Natsuhara K, Murayama N (2014) Shizen wo torikomu: Nisshihō ni yoru Raosu 3 chīki no yasei shokumotsu sesshu no hikaku (Taking in nature: comparing wild food ingestion in three regions of Laos through the diary method). In: Ikeguchi A, Sato R (eds) Neichā ando sosaethi kenkyu dai-3-kan: Shintai to seizon no bunka seitai (Nature and society research series, vol 3: cultural ecology of body and subsistence). Kaiseisha, pp 61–82 (in Japanese)

Ratsavong K, van Elsacker T, Doungvichit D, Siengsounthone L, Kounnavong S, Essink D (2020) Are dietary intake and nutritional status influenced by gender? The pattern of dietary intake in Lao PDR: a developing country. Nutr J 19(1):31, 1–16. https://doi.org/10.1186/s12937-020-00545-9

Sato R (2009) Hito no nariwai wa sei to shi ni dō kakawatte kita ka: Shinrin yakihata-min no raifu kōsu to jinkōshi (How have human occupations related to life and death? Life courses and population history of forest swidden agriculturalists). In: Ikeya K (ed) Chikyu kankyoshi kara no toi: Hito to shizen no kyosei to wa nani ka (Approaches to global environmental history: towards a harmony of nature and society). Iwanami Shoten, pp 60–78 (in Japanese)

Sato R (2014) Mori sumi no yakihata-min ga otona ni naru made: Echiopia shinrin yakihata-min no nariwai to seikatsushi (Until the forest-dwelling swidden agriculturalists grow up: occupations and life history of the Ethiopian forest swidden agriculturalists). In: Ikeguchi A, Sato R (eds) Neichā ando sosaethi kenkyu dai-3-kan: Shintai to seizon no bunka seitai (Nature and society research series, vol 3: cultural ecology of body and subsistence). Kaiseisha, pp 203–224 (in Japanese)

Sato R, Jiang H, Nishimoto F, Yokoyama S (2023) Food procurement strategies of swidden agriculturalists in central Laos: analysis of side dish ingredient records in food diaries. E-J GEO 18:309–323 (in Japanese). https://doi.org/10.4157/ejgeo.18.309

Sellen DW, Mace R (1997) Fertility and mode of subsistence: a phylogenetic analysis. Curr Anthropol 38:878–889. https://doi.org/10.1086/204677

Tomobe K (1988) A theory of peasant family economy and the Chayanovian theory: problems and perspectives. Mita Gakkai Zasshi (Mita J Econ) 81:505–529 (in Japanese). https://doi.org/10.14991/001.19881001-0145

Temporal-spatial Distribution of Swidden Agriculturists' Daily Activities in Central Laos

Hongwei Jiang, Ren'ya Sato, Satoshi Yokoyama, and Futoshi Nishimoto

Abstract The authors conducted three time-allocation surveys in a Mankhong ethnic minority hamlet in Central Laos, where swidden cultivation, fishing, hunting, and gathering are the main subsistence of the residents. In the surveys, three types of data were collected using the recall method to collect information on daily activities, a GPS logger and accelerometer to collect the spatial information, and energy consumption of daily activities. The results of the data analysis clarified the temporal-spatial pattern of daily activities, such as labor division between male and female adults in working-time allocation and activity space and its gender difference. Furthermore, from the balance of time allocation between various subsistence of swidden agriculturalists in Central Laos, where plantation development is progressing at a rapid pace, the authors tried to provide important clues for understanding the changes in land use and subsistence in the study area.

Keywords Time allocation · GPS logger · Accelerometer · Time geography · Central Laos · Swidden agriculturist

This chapter was a revised version of Jiang et al. (2023).

H. Jiang (✉)
National institute for Environmental Studies, Tsukuba, Japan
e-mail: jianghw@gmail.com

R. Sato
Graduate School of Humanities, Osaka University, Toyonaka, Japan
e-mail: rsato7788@gmail.com

S. Yokoyama
Graduate School of Environmental Studies, Nagoya University, Nagoya, Japan
e-mail: yokoyama.satoshi.i4@f.mail.nagoya-u.ac.jp

F. Nishimoto
Embassy of Japan in Laos, Vientiane, Laos

S. Yokoyama et al. (eds.), *Population Dynamics and Livelihood Changes of Small-Scale Societies in Laos*, International Perspectives in Geography 22,
https://doi.org/10.1007/978-981-96-2956-5_3

1 Introduction

This chapter aimed to clarify the time allocation, spatial distribution, and energy consumption of daily activities such as work related to subsistence activity, domestic work, and leisure of male and female adults, or those of adults and children by season, based on the data of time allocation survey and the records of its spatial information and energy consumption. Furthermore, this chapter revealed the contribution of residents by sex and age based on their work time of major subsistence (swidden agriculture), subsidiary subsistence (hunting, gathering, fishing), and domestic work. Basing on these results, the relation between time allocation and the change of land use and subsistence was discussed from perspective of time geography.

Time allocation is not only an important indicator for assessing the time spend on major and subsidiary subsistence, but also an unavoidable factor in understanding land use intensification in rural areas of developing countries. Pioneering studies in the field of land use intensification research included the land use intensification theory proposed mainly by Boserup (1965) in the 1960s and 1970s, former Soviet Union agricultural economist Chayanov's (1966) small farmer economics theory, and the spatiotemporal resource flow concept proposed from a time geography perspective by Carlstein (1982).

In the 1960s and 1970s, a group of researchers led by Boserup (1965) proposed a new land intensification theory in which population growth was the main driver of land use intensification. Two characteristics were emphasized in this land intensification model. The first is an increase in the frequency of land use (cropping). Second, while land productivity per unit area increases, productivity per unit labor time tends to decrease. The second feature implies that households have labor resources, but limited land resources for the production base, which means that they increase labor input excessively in order to increase yield. In other words, this phenomenon is a kind of mismatch between labor and land resources, which can lead to a decline in labor productivity. However, if a system of land rental and labor exchange exists in the community, a decline in labor productivity can be avoided.

On the other hand, in Chayanov's (1966) theory of small farmer economics, in a society without land use restrictions, commodity economy, and wage labor, land resources are available in proportion to the amount of labor (time) invested, so the over-input of labor to limited land resources discussed by Boserup above does not occur. According to Chayanov's theory, the ratio of the number of consumers to the number of workers in each household contributes to the amount of labor needed to meet the household's biological demand. In other words, the higher this ratio, the longer households have to work.

After analyzing the theories of Boserup (1965) and Chayanov (1966), time geographer Carlstein (1982: 296–297) proposed the idea of resource mobility. Resource mobility includes "space–time carrying capacity" (also known as land resource mobility) and "time resource mobility." "Space–time carrying capacity" is the carrying capacity of a given space and period. Factors that increase the demand for "space–time carrying capacity" include population growth, production taxation, land occupation by immigrants, and the demarcation of nature reserves. Factors associated with reduced demand for "space–time carrying capacity" include population emigration, importation of food and other products, introduction of land-saving crops (crops with high land productivity), and opening up of unused land. If factors that expand "space–time carrying capacity" are greater than the causes that contract it, the area's "space–time carrying capacity" will expand. Two possible methods to extend "space–time carrying capacity" involve expansion and intensification of the space used, and which of these methods predominates will depend on the forms of subsistence of the study area.

Carlstein (1982: 396–402) presents two ideas on the "time-resource mobility." The first is an increase in the time input in a particular activity, i.e., the intensification of labor time allocation. The other is an increase in total labor hours by increasing the total number of laborers in the household. The former certainly refers to increasing the workload burden. The latter refers to the strategy of increasing the population of laborers in the households. There are two possible methods for increasing the population of laborers: a short-term method of bringing adult laborers through marriage, and a long-term method of increasing the number of children through reproduction.

With the above theory on time allocation intensification, Carlstein (1982: 386–388) criticized Chayanov's small farmer economic theory on the two following points. First, there is surely no society such as the one Chayanov suppose, with no commodification of products; and second, changes to household structure occur owing to not only simple biological life cycles (reproduction of children) but also to external factors (replenishment of the labor force).

Today's rural societies in developing countries are vastly different from the Russian rural villages that Chayanov observed and analyzed at the start of the twentieth century, and from the developing country's rural societies of the 1960s–1970s and earlier that Boserup and Carlstein used as analysis material. Population growth due to demographic and health transitions,[1] development through external initiatives, and demarcations for nature reserves are some of the major constraints on the use of land resources that form the foundation of survival for rural communities in developing countries. In addition, the penetration of the commodity economy has contributed to the increase of land rent and trade, wage labor, and the establishment of an agroforestry product trade system in local communities. Under these conditions, the only thing that has not changed is people's individual time resources.

People live their lives by allocating their limited 24-hour day to work, sleep, eat, and rest, and other activities. Clarifying the time allocation of these life activities is essential in understanding the current use of land resources that forms the basis of survival. This will allow us to discover the balance and imbalance of time and land resource allocation, providing important knowledge for predicting future trends in rural land use changes in developing countries. Against this background, the

[1] Health transition is a system concept that shows the gradual structural transitions in a country's health issues, based on the transition of population structure, and changes in healthcare systems and socioeconomic structures, based on the mutual effects of changes in population and disease patterns, in a particular time period (Japan Association for International Health, https://jaih.jp/document/term/, accessed January 4, 2023).

purpose of this chapter is to clarify the time and spatial allocation of daily activities of swidden agriculturalists in central Laos, where plantation development through external initiatives are rapidly advancing.

2 Study Area

The study area—AN hamlet is located in Xepon District, Savannakhet Province, central Laos, and is inhabited by the Mankhong people, a Mon-Khmer ethnic minority group (Fig. 1). The hamlet is located on the hill area along No. 9 National Highway, which connects Kaison Phomvihane, the capital of Savannakhet Province, to the border checkpoint with Vietnam, at an elevation of about 210 m. It is about 20 km from the town area of Xepon District, but as the Xepon River lies between the hamlet and Highway No. 9, the hamlet's transport access is by no means convenient. Xepon River as an obstacle affects the daily activities of the residents including use of transportation methods, such as motorbikes or cars, and their access to all types of infrastructure facilities. In terms of modernization, the hamlet is clearly lagging behind the adjacent AG hamlet[2] on the opposite shore. The AN hamlet received electricity in 2019, and then water wells were finally installed in 2021 through trial and error, thereby resolving their potable water issues.

In Xepon District where AN hamlet is located, large-scale rubber, acacia, and eucalyptus (raw materials for paper manufacturing) afforestation began in the mid-2000s with capital from China and Vietnam, and the secondary growth forests that were traditionally used for swidden agriculture were rapidly converted into these plantations. These plantations have not directly affected AN hamlet's swidden land, but migration from plantation-adjacent villages is certainly pressing in on the vicinity of AN hamlet as people search for swidden land. In addition, government initiatives have caused people to migrate from even more remote mountain hamlets to hamlets like AN that are close to highways, contributing to the population growth of AN hamlet. As a result, 35 people from 9 households migrated to the AN hamlet during the 5 year period from 2015 to 2020, increasing the population. This migrant population represented about a quarter of the population of the AN hamlet in 2020. This sudden population growth had resulted the reduction of per capita land resources for swidden cultivation in AN hamlet.

As of February 2020, there were 149 people in 29 households living in the hamlet. Most residents' major subsistence was producing sticky rice through swidden cultivation while practicing hunting, gathering, and fishing as subsidiary subsistence. Hamlets with easy access to the main road of the National Highway No. 9 have introduced the cultivation of cash crops such as cassava and bananas, but in this hamlet with no easy access to transportation, only two households have cultivated bananas on a total of 1 ha (interview with hamlet head; September 2022). Furthermore, many residents in surrounding hamlets are employed at the rubber plantations

[2] AN hamlet and AG hamlet belong to the same administrative village.

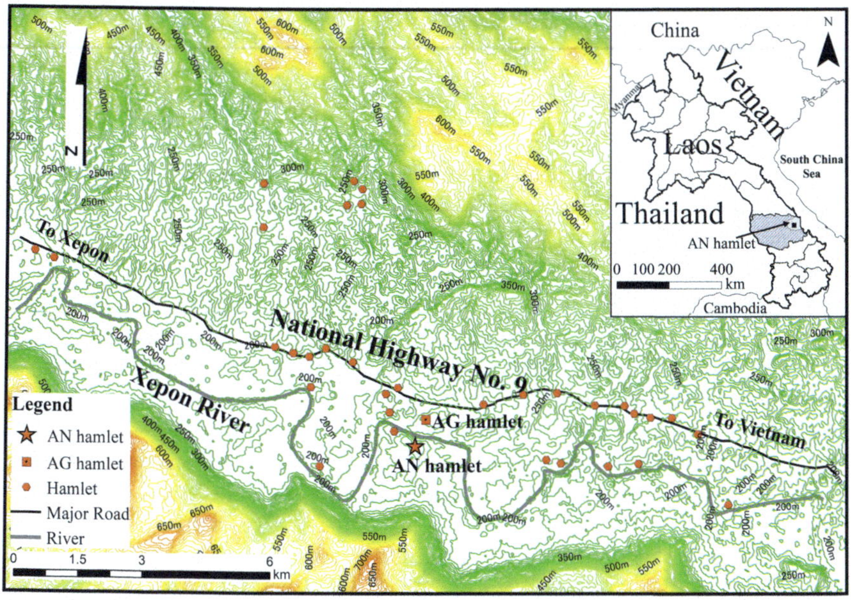

Fig. 1 The geographical location of study area. *Source* SRTM data taken in February 11–22, 2000. *Note* AN hamlet located in the south side of No. 9 National Highway from the Savannakhet provincial capital city of Kaison Phomvihane to the border between Vietnam and Laos

operated by Vietnamese company, but almost no one pursues this type of employed labor in AN hamlet.

AN hamlet's swidden agriculture is practiced in accordance with the precipitation changes of the dry and rainy monsoon season. According to the Xepon weather station's precipitations records from 1988 to 2018, the annual mean rainfall is 1791 ± 423 mm,[3] of which approximately 90% is concentrated in the rainy season from April to October. Accordingly, February to March each year, around the end of the dry season, is the period for slashing and burning for swidden cultivation. When the rains come around mid-April to early May, the dry-land rice is sown. The main agricultural work done during the three to four months from sowing to harvest is weeding. Harvest time depends on the variety of sticky rice, with early varieties harvested in late September and late varieties harvested in late October or early November. Residents of this hamlet do their yearly subsistence activities to fit with the seasonal changes of swidden operations.

[3] According to data from the meteorological station in Xepon District, Savannakhet Province.

3 Research Materials and Methods

3.1 Survey Methods for Time Allocation Studies

Time allocation studies have been used as a method for quantifying periods of time not only in geography but also in many other fields such as cultural anthropology, agricultural economics, development studies, domestic science, sociology, and labor science (Noma 1988). In the above research fields, time allocation studies have proposed survey methods to measure various types of life activity time in a quantitative or semi-quantitative manner. As Noma (1988) summarized, there are a variety of methods for collecting time allocation data, including the "interview method," "participant observation method," "spot check method,"[4] and "self-administration method." Each of these data collection methods has its advantages and disadvantages. For example, participant observation is a method in which the researcher participates in daily activities with the study participants and records participants' time allocation in their daily life activities through direct observation. This type of method enables a more precise understanding of participants' time allocation, but only a limited number of participants can be observed, and the effectiveness of this method is extremely limited for acquiring a full view of the time allocation of a study population.

However, while these time allocation survey methods have enabled researchers to grasp information about places of action, sufficient information could not be obtained about the space (positioning) of the places wherein people acted. This lack of activity spatial information implied that conventional survey methods were insufficient for researching and analyzing interactions between humans and their living environments and the effects of the organization of living spaces on time allocation (e.g., analysis of travel time from hamlet housing to shifting fields.). In this context, from the 1970s to 1980s, Swedish geographer Torsten Hägerstrand and his collaborators proposed the concept of time geography (Castree et al. 2013: 516–517) linking activity time and space. They represented each type of constraints that could affect human activity with a time–space prism.[5]

Carlstein (1982) used the prism to investigate studies on human groups' time allocation and land use in pre-industrialized societies that practiced hunting, gathering, swidden agriculture, and irrigation farming and compiled the features of these societies' use of space and time from the perspective of time geography. Recent advances in spatial analysis and visualization technologies have made it possible to conduct time geography research to analyze human activity patterns by incorporating large amounts of data on human movements, the socioeconomic status

[4] The spot-check is a method of estimating the time allocation of a target activity based on the observed frequency over a long period of time (e.g., one year) at a randomly determined date and time (Moji 1985; Ohtsuka 2011: 69–81).

[5] Prism is a diagram that shows the spatio-temporal range that can be reached by an individual's activity path, and is shaped by the speed of movement and the absolute location of the starting point (Thrift 1977; Noma 1988).

of survey targets, and public statistical data into a geographic information system (GIS) (Kwan and Lee 2004). Regarding the rural research in developing countries, Nishimura's research group provided good example of time geography study on paddy farmers in the outskirts of Vientiane, Laos. They collected spatial information of activity by portable GPS loggers, activity type and activities' spatiotemporal paths during the survey period by interview. Based on these data, they created activity time–space prisms of survey participants and discussed the spatiotemporal features of target population's daily activities (Nishimura et al. 2010).

Several time allocation studies have also been developed in Laos, albeit with different research objectives. Among them, Phouyyavong et al. (2020) visited 133 residents in a village in northern Laos where a combination subsistence including swidden agriculture and cash cropping is practiced, revealed individuals' time allocation and subsistence portfolio by collected annual and monthly working hours, and stressed the importance of swidden agriculture in sustaining livelihoods. Although these data collecting methods of time allocation are less precise than the time geography and human ecology methods mentioned in this chapter, such as observation and daily interview, they are important for capturing an overall picture of time allocation by subsistence in a target population. On the other hand, Nishimura et al. (2010), Yamauchi (2018), and Jiang et al. (2021) used participant observation and daily interview to investigate the time allocation of paddy farmers in Vientiane and western Savannakhet provinces. Yamauchi (2018) identified the time allocated to various subsistence activities and estimated each individual's energy expenditure by considering indicators such as time allocation, physical activity level, height, weight, and gender for each activity. However, Nishimura et al. (2010) and Jiang et al. (2021) focused on spatial information and conducted time geographic analysis, but the emphasis of their analysis differed. Nishimura et al. (2010) used portable GPS data to analyze the spatio-temporal prism of the subjects' daily activities, while the Jiang et al. (2021) focused on the analysis of directional distribution ellipses with portable GPS data, as shown in this chapter. Given the almost complete lack of time geography analyses of swidden agriculturist communities in Laos, especially those less affected by development, the results of this chapter can provide important information for understanding the spatio-temporal use of pre-development Lao swidden agriculturist and for predicating the trend of future change.

Based on these previous studies, this study devised a "recall method" for grasping the overall situation of target population's time allocation. This "recall method" made it possible to collect the information about activity type and its time of a large number of people in a short period of time. This "recall method" has two characteristics. First, target population's activities are ascertained through long-term observation in the field, and these activities are categorized and coded. Second, unlike the conventional "interview method," the main activities are identified and activity codes are recorded in hours rather than minutes. Compared to the conventional "interview method," this method reduces the time required for interviews, reduces the burden on survey targets, and make it possible to collect data from a larger number of people. Although the accuracy of individual data will be reduced, the increase in the number of survey targets will make it easier to get an overall picture of the target population.

In addition to the "recall method," this study incorporated GPS loggers to collect spatial information on activity and accelerometers to assess the energy expended in the physical activity of targets. Using time stamp from GPS loggers and accelerometer recordings, the spatial information of activities and their energy consumption were combined to enable the evaluation of the spatial distribution of targets' energy consumption and work intensity.

3.2 Data Collection

In order to examine seasonal changes in activity, a time allocation survey was conducted with AN residents on three occasions during the dry and rainy seasons. The surveys were conducted in June 2016 (rainy season), February 2017 (dry season), and February 2020 (dry season). The participants in the June 2016 and February 2017 surveys were all adults (16 years and older, defined as "adults" below); the February 2020 survey, however, included study participants who were minors aged 6–15 ("minors" below) (Table 1). In the time allocation survey, we collected three types of activity data: recordings of hourly activity content during the daytime (6:00–19:00) using the recall method (Table 2); spatial information on activities (collected at 12 s intervals) and swidden positioning information using GPS loggers (M241, Holux Technology, Inc.)[6]; and finally, measurements of participants' physical activity intensity (energy expenditure) using accelerometers (Lifecorder EX, Suzuken Ltd.).[7,8] In all surveys, we explained the survey process and considerations regarding personal information to survey participants or their guardians, and after receiving consent, we asked them to wear a GPS logger and accelerometer (Fig. 2) and participate in an interview on their activities.

We also used the spatial information collected with the GPS loggers to assess participants' overall spatial distribution of activity (relative positions to the geometric center of hamlet) and the range and direction of individual activities. We assessed the overall spatial distribution of activity by setting 10×10, 500 m grids starting at the geometric center point of the hamlet while counting the number of GPS recordings made in each grid during the survey period. To assess individual activity ranges and directionality, we used the area of standard deviation ellipses (hereafter referred to

[6] The GPS logger (M-241, Holux Technology) is a lightweight GPS receiver. According to the manufacturer's report, the horizontal error for DGPS (EGNOS/WAAS/MSAS) reception is within 2.2 m and the vertical error is within 5 m. Considering the accuracy of the survey and the burden on the participant (light weight), the M-241 was used in the survey related to this study.

[7] The accelerometer used to measure energy consumption, Lifecoder EX (Suzuken), is a weighing instrument that estimates the energy expended in physical activity based on the frequency and intensity of human body (center of gravity) movements. For a detailed method of estimating energy consumption, see Kumahara et al. (2004).

[8] Accelerometer measurements were not taken during the February 2020 survey due to instrument failure.

Table 1 Age and sex composition of participants in time allocation survey

Season and period	Age	Sex	Number
June 2016, rainy season (5 days)	Adult (>15 years)	Male	13
		Female	12
February 2017, dry season (3 days)	Adult (>15 years)	Male	6
		Female	7
February 2020, dry season (4 days)	Adult (>15 years)	Male	19
		Female	21
	Minors (age 6–15)	Male	5
		Female	4

Source Authors' survey

as ellipses) calculated using the spatial statistics tool "Direction Distribution Analysis"[9,10] in ArcGIS 10.5, it's direction. For activity ranges and directional ellipses, we used a 2-SDE (Standard Deviation Ellipse; Hirsch et al. 2014; Patterson and Farber 2015) capable of covering 98% of each survey participant's GPS recordings. In addition, we used the GPS loggers to trace each participant's swidden boundary lines and made a swidden distribution map corresponding to the survey season. To examine the spatial distribution of participants' physical activity intensity, we used time stamps recorded by the accelerometers and GPS loggers to add location information to the physical activity intensity that was recorded. By analyzing this data, we grasped the spatial distribution of participants' energy expenditure. We used SPSS 25.0 (IBM Ltd.) for statistical analysis of the data.[11]

We estimated physical activity intensity by using the accelerometer recordings to convert physical activity into the assessment indicator of Metabolic Equivalents (METs)[12] (Kumahara et al. 2004). In this study, we considered accelerometer activity recordings of 2 and above (MET $\geq$ 2.3) as moderate to vigorous physical activity.

[9] Direction distribution analysis is an ArcGIS (ESRI) tool for analyzing spatial distribution patterns of features. The features in this paper are time-series GPS logger records of each subject's activity. For more information on direction distribution analysis, see https://pro.arcgis.com/ja/pro-app/2.9/tool-reference/spatial-statistics/directional-distribution.htm (Accessed 6th April, 2023).

[10] The standard deviation ellipse of the direction distribution reflects the activity space pattern of each participant during the survey period. 1-SDE (Standard Deviation Ellipse), 2-SDE and 3-SDE can cover 63, 98 and 99% of the activity records, respectively. Since the 1-SDE may underestimate the participant's activity range and the 3-SDE may overestimate the activity range, the area of the 2-SDE was used in this chapter to evaluate the area of the participant's activity space.

[11] For reasons of space limitations, the authors generally do not present the computational process using statistical software in the paper. If readers have questions or concerns, the authors are responsible for providing the raw data and the computational records described in the text.

[12] Metabolic equivalents (METs) are defined as the caloric consumption of an active individual compared with the resting basal metabolic rate at rest (Boccalandro 2010).

Table 2 Classification of activity as used in activity survey conducted by the recall method

Activity category	Details
1. Shifting and paddy fields work	Sowing, rice planting, weeding, spraying pesticide, fertilizing, plowing fields, managing water, and harvesting
2. Rubber plantation	Tapping, weeding, and collecting latex, among other things
3. Home gardening	Sowing, weeding, spraying pesticide, fertilizing, harvesting, and managing water
4. Fishing	Catching fish, shrimp, crabs, etc. from rivers and ponds
5. Raising domestic livestock	Taking care of cattle, water buffalo, goats, pigs, and domestic fowl or cutting grass to feed them
6. Gathering forest products and insects	Gathering insects, frogs and tadpoles, ant eggs, pond snails, edible wild grasses, bamboo shoots, mushrooms, rattan, and other edible plants
7. Hunting activities	Large animals (boar, deer, etc.), bamboo rats, squirrels, birds, mice, and snakes, among other animals
8. Weaving	Weaving-related activities
9. Commercial activities	Running stores and peddling
10. Housework	Laundry, dishwashing, childcare, housecleaning, drawing water, making tools such as brooms, cooking
11. Eating	Breakfast, lunch, dinner, and snacks
12. Leisure, etc.	Visiting neighbors, having small conversations, playing games, watching TV, taking breaks, napping
13. Personal hygiene	Shower, brush teeth, wash face, etc.
14. Social life	Meetings, visiting people who are unwell, temple prayer activities
15. Employed labor	Factory work or employment at rubber plantations and other jobs through employment with individuals or companies
16. Other	Activities other than the above

Source Authors' survey

4 Results

This chapter describes the results of the survey conducted using the methods described in Sect. 3.2.

Table 3[13] show the time allocation of adult men and women, respectively, as collected through the recall method in June 2016 (rainy season). For the average

[13] Since there was only work related to shifting field, the term "shifting filed and paddy field works" in Table 2 was changed to "shifting field work" for convenience. Tables 4 and 5 were same.

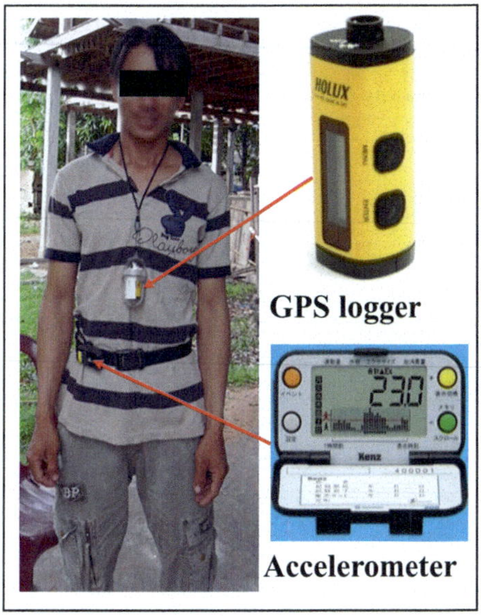

Fig. 2 Activity survey participant wearing GPS logger and accelerometer

total hours and standard deviation of subsistence work, i.e., home gardening, fishing, livestock rearing, forest products and insects gathering, hunting of wild animal, weaving, commercial work, adult men and women worked 4.3 ± 1.2 and 4.8 ± 1.6 h respectively. Comparing the time allocation of adult men and women for each activity item, there was no significant difference between males and females during rainy season.

Table 4 show the time allocation of adult men and women, respectively, as collected through the recall method in February 2020 (dry season). For the average total hours and standard deviation of subsistence work, adult men and women worked 3.1 ± 1.6 and 2.9 ± 1.2 h respectively. Comparing the time allocation of adult men and women for each activity item, adult men spent significantly longer hours than women on fishing. Meanwhile, adult women spent significantly longer working hours on gathering, weaving, and housework than men.

Table 5 show the time allocation of male and female minors, respectively, as collected through the recall method in February 2020 (dry season). For the average total hours and standard deviation of subsistence work, male and female minors worked 2.4 ± 0.7 and 2.6 ± 0.5 h respectively. Female minors spent more hours than male minors on housework.

Figure 3 presents a 5 km grid map created based on the activity spatial information recordings of adult men and women that were gathered in June 2016 (rainy season), February 2017 (dry season), and February 2020 (dry season). The secondary grid in this map shows a range of 500 m. The depth of color in the secondary grid represents the frequency of activity recordings. In both the rainy and dry seasons, the activity range of both men and women was concentrated within 2.5 km from the geometric center of the hamlet. It was observed that the activity area in February

Table 3 Time allocation and differences between adult (>15 years) men and women in the rainy season (June 2016). *Unit* hour

Age/activity category	Male (13)		Female (12)	
	Mean	SD	Mean	SD
Age	32.2	12.3	33.0	11.5
Total subsistence work time	**4.3**	**1.2**	**4.8**	**1.6**
Shifting fields work[13]	1.2	1.4	1.3	1.5
Gardening work	0.9	1.1	1.7	1.5
Fishing	0.7	0.9	0.9	1.0
Raising domestic livestock	0.7	1.6	0.6	1.2
Gathering forest products and insects	0.6	0.8	0.3	0.5
Hunting wild animals, etc.	0.2	0.5	0.0	0.1
Weaving	0.0	0.0	0.0	0.0
Commercial activities	0.0	0.0	0.0	0.0
Housework	**2.9**	**1.2**	**2.7**	**0.9**
Leisure	**5.7**	**0.6**	**5.6**	**1.0**
Employed labor	**0.0**	**0.0**	**0.0**	**0.0**
Other	**0.1**	**0.2**	**0.0**	**0.0**

Source Authors' survey
Note No significant difference was found in the time allocation to any activities ($p < 0.05$, Mann–Whitney U test)

2020 was considerably larger than in February 2017 during the same dry season. The difference in the number of informants may have made the difference (Table 1).

Figure 4 shows some survey participants' activity ranges (2-SDE) during the survey period, calculated using spatial statistics tool "Direction Distribution" in ArcGIS. The area of each ellipse represents each participant's activity area, while the direction of the ellipse shows the direction of each participant's activities. Table 6 shows the adult men and women's or male and female minors' area of activity space obtained from survey participants in the three surveys of June 2016, February 2017, and February 2020. In order of survey season, the area of activity space for adult men was 4.9 ± 6.9, 2.5 ± 1.5, and 5.7 ± 5.8 km^2, and that for adult women, 1.5 ± 1.5, 0.1 ± 0.1, and 2.5 ± 4.4 km^2. The mean activity area for male and female minors (age ≤ 15) in February 2020 was 7.9 ± 10.3 and 0.6 ± 0.8 km^2, respectively. A general tendency was observed that men's (including minors) activity area was more extensive than that of women, but the difference was only significant in February 2020 (dry season). Comparing male adults and minors, and female counterpart in the February 2020 survey, no significant differences were found in either case.

Figure 5 shows the spatial distribution of adult men and women's physical activity intensity in June 2016 (rainy season) and February 2017 (dry season). In the rainy season, about 80% of women's light and moderate to vigorous physical activity was concentrated within a range of 0.5 km from the hamlet geometric center, while a

Table 4 Time allocation and differences between adult (>15 years) men and women in the dry season (February 2020). *Unit* hour

Age/activity category	Male (19)		Female (21)	
	Mean	SD	Mean	SD
Age	32.0	14.6	37.1	16.3
Total subsistence work time	**3.1**	**0.6**	**2.9**	**1.2**
Shifting fields work[13]	1.7	0.9	1.1	1.1
Gardening work	0.0	0.0	0.1	0.2
Fishing***	1.0	0.5	0.1	0.2
Raising domestic livestock	0.1	0.2	0.0	0.0
Gathering forest products and insects***	0.2	0.3	1.0	0.7
Hunting wild animals, etc.	0.1	0.2	0.0	0.0
Weaving*	0.0	0.0	0.6	1.0
Commercial activities	0.0	0.0	0.0	0.0
Housework* **	**0.9**	**0.5**	**2.3**	**0.8**
Leisure* **	**8.9**	**0.5**	**7.8**	**1.0**
Employed labor	**0.0**	**0.0**	**0.0**	**0.0**
Other	**0.1**	**0.2**	**0.0**	**0.0**

Source Authors' survey
Note: * $p < 0.05$; *** $p < 0.001$, Mann–Whitney U test

Table 5 Time allocation and differences between adult (≤15 years) male and female minors in the dry season (February 2020)

Age/activity category	Male (5)		Female (4)	
	Mean	SD	Mean	SD
Age	**9.4**	**3.0**	**10.8**	**2.2**
Total subsistence work time	2.4	0.7	2.6	0.5
Shifting fields work[13]	0.3	0.6	0.0	0.0
Gardening work	0.0	0.0	0.0	0.0
Fishing	0.5	0.5	0.4	0.5
Raising domestic livestock	0.2	0.3	0.0	0.0
Gathering forest products and insects	1.4	0.5	1.9	0.1
Hunting wild animals, etc.	0.1	0.1	0.0	0.0
Weaving	0.0	0.0	0.3	0.3
Commercial activities	0.0	0.0	0.0	0.0
Housework*	**0.3**	**0.1**	**2.6**	**0.8**
Leisure	**10.1**	**1.4**	**7.8**	**1.2**
Employed labor	**0.3**	**0.7**	**0.0**	**0.0**
Other	**0.0**	**0.0**	**0.0**	**0.0**

Source Authors' survey
Note * $p < 0.05$, Mann–Whitney U test

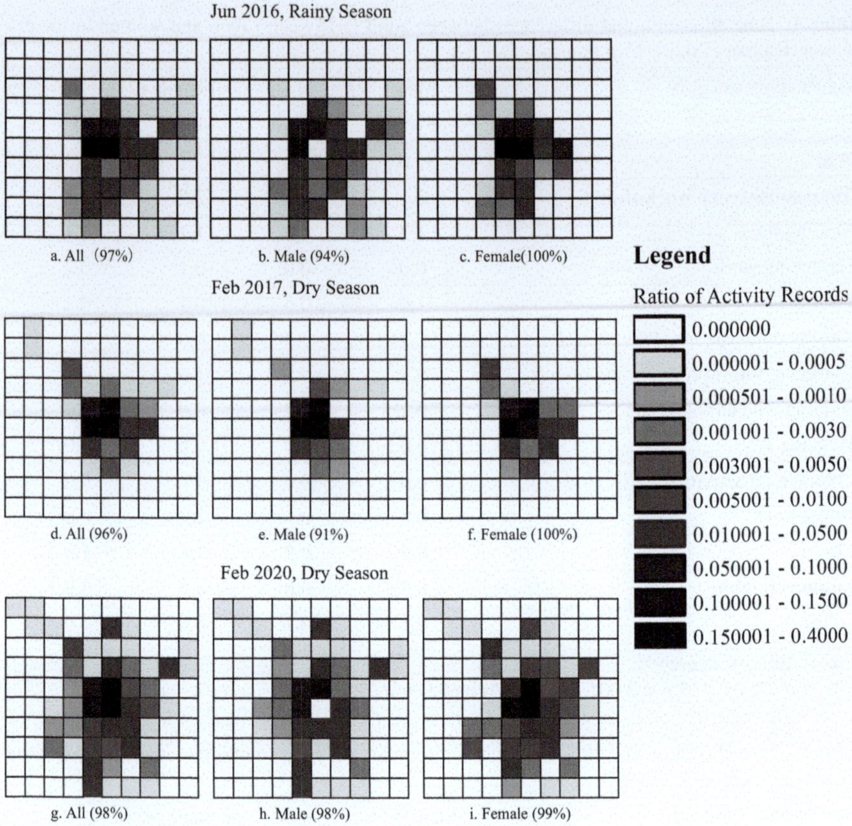

Fig. 3 The spatial distribution of adult residents' daytime activities in AN hamlet. *Source* Authors' survey. *Note* Panels a–i was the first-order 5 km grid maps created from the geometrical center point of the hamlet. The secondary grid in each map represents a 500 × 500 m area. The colour density of each secondary grid represents the frequency of the survey participants' activities recorded during the survey period. Panels a–c are the spatial distributions of all, male, and female participants' activities in June 2016. Panels d–f are those in February 2017; and panels g–i are those in February 2020. The percentages in brackets indicate the proportion of activity recorded within the 5 km grid

little under 40% of men's moderate to vigorous physical activity was conducted 0.5–1.5 km from the geometric center of the hamlet. In the dry season, no major differences were observed in the spatial distribution of male and female physical activity, but compared with the rainy season, male physical activity during dry season was more concentrated within a 0.5 km range of the village.

Figure 6 show the relative position between each household's residence and the location of their shifting field in 2016. The red lines on the map indicated the distance from the residence to the geometric center point of the shifting field. The mean and standard deviation of the distance for 2016 was 0.9 ± 0.4 km (n = 48, median value = 1.1 km; minimum vale = 0.2 km; maximum value = 1.8 km). The number of

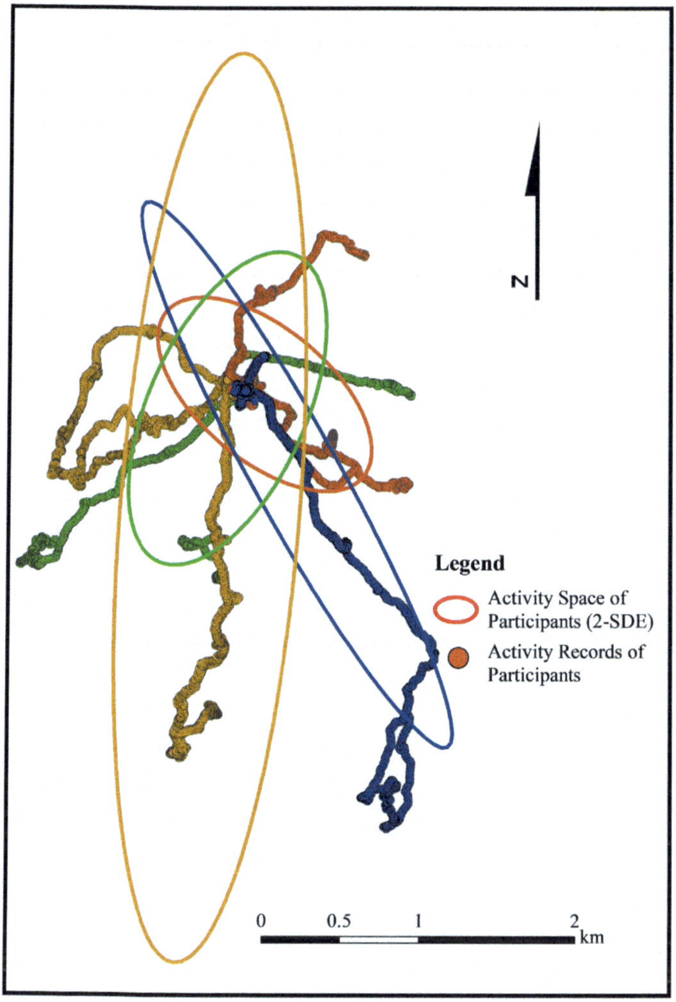

Fig. 4 Activity spaces created by recordings of participants' activity spatial records. *Source* Authors' survey

shifting fields closer than the average distance of 0.9 km was 23. The average area of shifting fields per household and the standard deviation were 1.4 ± 0.7 ha

Table 6 Difference in activity spaces between male and female adults/minors in the rainy and dry seasons

survey season	Item	Age-sex group			
June 2016 (rainy season)	Sex (number of participants)	Adult (>15 years)			
		Male (13)		Female (12)	
		Mean	SD	Mean	SD
	Age	32.2	12.3	33.0	11.5
	Area of activity space (km^2)	4.9	6.9	1.5	1.5
February 2017 (dry season)	Sex (number of participants)	Adult (>15 years)			
		Male (6)		Female (7)	
		Mean	SD	Mean	SD
	Age	32.0	12.0	34.9	8.2
	Area of activity space (km^2)	2.5	5.7	0.1	0.1
February 2020 (dry season)	Sex (number of participants)	Adult (>15 years)			
		Male (19)		Female (21)	
		Mean	SD	Mean	SD
	Age	32.0	14.6	37.1	16.3
	Area of activity space (km^2)*	5.7	5.8	2.5	4.4
	Sex (number of participants)	Minors (6–15)			
		Male (5)		Female (4)	
		Mean	SD	Mean	SD
	Age	9.4	3.0	10.8	2.2
	Area of activity space (km^2)	7.9	10.9	0.6	0.8

Source Authors' survey
Note *$p < 0.05$, Mann–Whitney U test

5 Discussion

5.1 AN Hamlet Land Use from the Perspective of Time Allocation

5.1.1 Limitation of AN Hamlet Residents' Working Hours

The combined subsistence and housework working hours of AN hamlet's adult men and women was 7.2 and 7.4 h, respectively, in the rainy season and 4.1 and 5.2 h, respectively, in the dry season (Tables 3 and 4). The working hours of male and female minors during the dry season were 3.0 and 5.2 h, respectively. The daily working hours of adult men and women during the rainy season, the busy farming season, were more than one hour longer than the 6.0 working hours of swidden agriculturist (Machiguenga, Upper Amazon) summarized by Minge-Klevana (1980). According

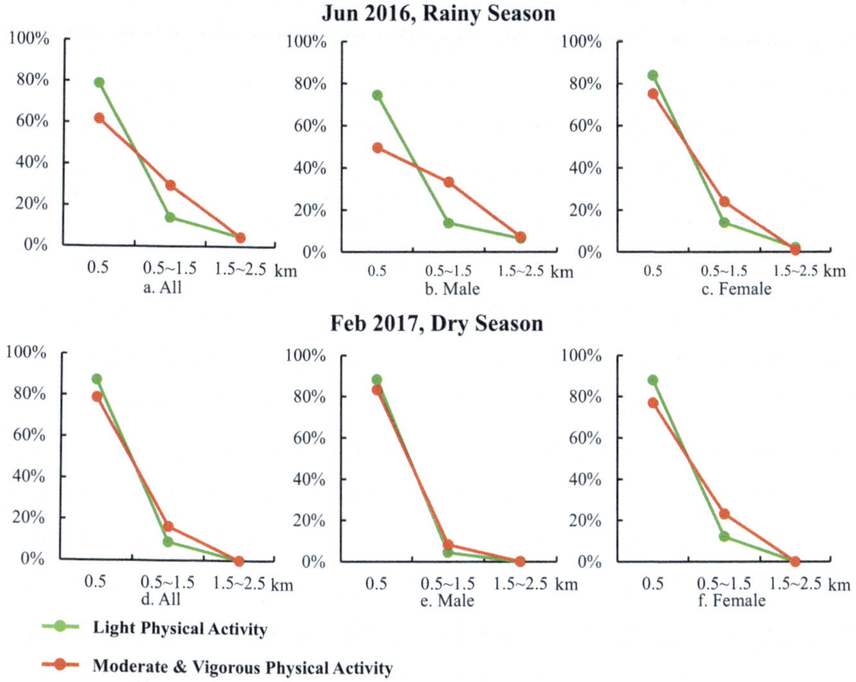

Fig. 5 Spatial distribution by intensity of adult men and women's physical activity in dry and rainy seasons. *Source* Authors' survey. *Note* The green and red lines indicate the percentage of light and moderate/vigorous physical activities, respectively, in 0.5, 0.5–1.5 and 1.5–2.5 km grids from the geometric center of AN hamlet. The y-axis represents the percentage of light or moderate and vigorous—physical activity records, while x-axis represents the distance from the geometric center of AN hamlet

to the authors' participant observations, the actual working hours in AN hamlet are likely to be even longer than this, since they had already started preparing meals and raising livestock around 4:00 am, in addition to the survey hours of 6:00 am to 7:00 pm. Another noteworthy point was the length of working hours of minors during the dry season. Male and female minors spent nearly the same level of working hours on subsistence activities as did adults in the same period. In particular, female minors' housework hours were at the same level as that of adult women.

5.1.2 The Need to Expand AN Hamlet's "Space–Time Carrying Capacity"

Referring the requirements for expanding "space–time carrying capacity" proposed by Carlstein (1982: 296–297) above, the AN hamlet had already experienced both population immigration and growth, and there were concerns that land occupation will occur owing to agricultural development in the future. There has been notable

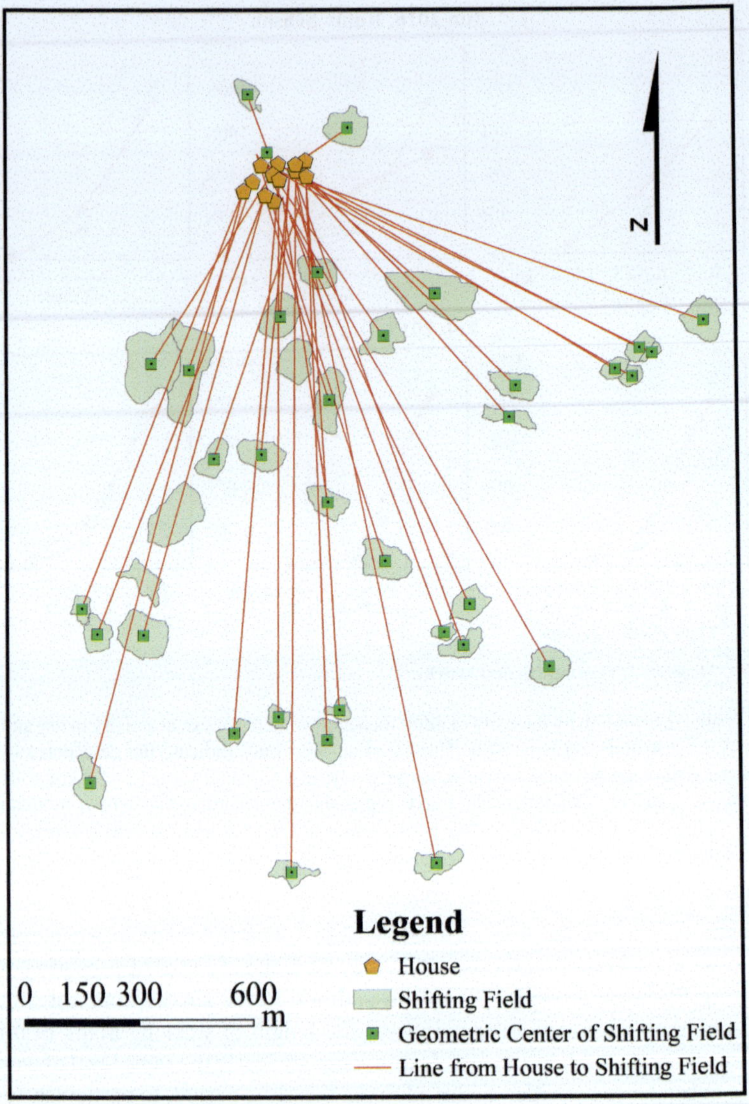

Fig. 6 Positional relationship between house and shifting fields for each household (based on 2016 shifting fields). *Source* Authors' survey

immigration between 2015 and 2020. According to the authors' interviews, AN hamlet grew by 14 households in those five years. Excluding the five households that grew as branch families as children got married, the other nine households came from even more remote mountain areas in search of the infrastructural convenience of being alongside a highway or that of support from relatives. The population surged from 89 in December 2015 to 149 in February 2020.

Calculating based on the shifting field area that the authors measured in 2016, that area per person was merely 0.3 ha. The rice yield (unhulled rice) weighed in 2011 in the Xepon area was 1.5 t/ha, so the per capita rice production in the study area can be estimated to be about 450 kg/person/year. The consumption amount per adult man measured by the authors in the 2017 dry season was about 422 kg (unhulled rice); therefore, we may assume that AN hamlet residents' food supply is just barely in balance. According to the interviews, some households were compelled to borrow from relatives or sell off domestic livestock, forest products that they had gathered (bamboo shoots), and wild animals that they had hunted, to buy rice owing to shortages around July and August, before the shifting harvest begins. If crops fail due to flooding or animal damage, food shortages are likely to occur. Based on the above perspective on food supply, AN hamlet may need to expand its space–time carrying capacity.

Additionally, in AG hamlet, located on the other side of the Xepon River, the immigration of the residents, who lost their shifting fields to plantation development was widely observed. In the vicinity of AN hamlet, the shifting fields operated by the immigrants of AG hamlet is increasing, and there is concern that there will be a shortage of land for swidden agriculture in AN hamlet in the future. However, expanding the space–time carrying capacity by expanding the swidden area would be challenging, as it would lead to greater travel distances between the hamlet and fields.

5.2 Discussion of Work Time Allocation and the Direction of Land/Time Resource Intensification

Assuming that the swidden agriculture pattern of subsistence is maintained, it would be difficult to apply Boserup's land use intensification model to time allocation in AN hamlet. As mentioned earlier, in Boserup's (1965) model, additional labor (time) must be invested to maintain or increase the productivity of the land. Considering the current state of AN hamlet adults' time allocation, it seems possible to increase the number of working hours further by cutting down break periods to some extent. In the period from 10:00 to 15:00 during the rainy season, adult men and women spend an average of 1.9 and 1.7 h, respectively, on leisure and other activities. However, given the high temperatures during this time period, further increases will not be easy. Additionally, there are limits to the possibility of increasing work hours by increasing the amount of labor power. As shown in Table 5, during the dry season, even minors spent 2.7 h per day for male and 5.2 h per day for female on subsistence activities and housework. In the rainy season, the amount of time spent on forest products gathering and fishing could be further increased. In fact, Sato et al. (2025) analyzed a food diary survey conducted in the same hamlet and found that in households with children aged 10 and older, the frequency of children contributing to food through activities such as gathering and fishing reached around 30%. Taking children's growth processes and physical strength into consideration, we can surmise that investing more working hours would be arduous for them.

Meanwhile, it would be difficult to explain AN hamlet's working hours and land use situation based on Chayanov's small farmer economic theory. According to Chayanov's theory, the higher the ratio of a household's consumers to its laborers, the larger its farming land area will be, and the number of working hours invested will also increase in proportion. In the case of AN hamlet, this would imply investing more working hours in swidden agriculture. Looking at the actual work hours of activity items involved in the subsistence activities of AN hamlet, swidden agriculture activities involved the greatest number of hours for both adult men and women. As the group with the greatest number of hours in swidden agriculture, adult men's number of hours in that were 1.2 and 1.7 h in the rainy season and dry season, respectively, accounting for one third to one half of their subsistence working hours of 4.3 and 3.1 h, respectively (Tables 3 and 4). Considering the perspective of work hours, if they reduced the number of work hours in other subsistence and increased the shifting field area and work hours, perhaps they could improve their food supply situation.

This raises the question of why the residents did not do so. Aside from land resource constraints, another factor, called the labor time bottleneck, may also explain to some extent. As the slashing of fallow forest for swidden agriculture is limited to a fixed period, the work hour resources that can be invested during this fixed period have a major effect on the actual area of cultivation. AN hamlet residents do their swidden slashing without the use of modern tools such as chainsaws; thus, the work hour resources that each household possesses at the end of the dry season becomes an important limitation on the swidden area that they can cultivate. Male and female minors are involved in jobs related to swidden agriculture, but in the authors' observations, their swidden slashing activities were minimal. The contribution of minors in swidden agriculture operations was mainly in weeding. In other words, the contribution of minors to the expansion of the shifting field area is limited. Unless high-yielding dry rice varieties and cultivation techniques are introduced, or a market-oriented subsistence transition occurs, it will be difficult to increase food production in the traditional swidden agriculture subsistence system.

The time resource mobility proposed by Carlstein could explain AN hamlet's situation. Considering the perspective of time resource intensification, the diversity of AN hamlet's subsistence work hours may be the intensification of work hour resource allocation. Owing to the limitations on the shifting field area, AN hamlet residents do not invest any more work hours in swidden agriculture than necessary; instead, they invest work hours in subsidiary subsistence such as hunting wild animals, gathering forest products and insects, and fishing. Adult men and women spent 1.5 and 1.2 h, respectively, on these subsistence during the rainy season (June 2016, Table 3) and 1.3 and 1.1 h, respectively, during the dry season (February 2020, Table 4). In both cases, they were investing a similar amount of working hours as they were to swidden agriculture activities in the same period. In contrast, male and female minors spent 2.0 and 2.3 h, respectively, on subsidiary subsistence activities during the dry season (Table 5), accounting for over 74% of their subsistence work hours. The products of these subsidiary subsistence not only provided to supplementary food for residents but also contributed to a source of cash income by selling of some products. That is then set aside for purchasing staple foods and everyday supplies. Based on this, they contribute to a stable food supply. This allocation of time to the subsistence that can obtain higher yields could be considered a type of time resource intensification.

Meanwhile, a trend toward allocating time to high-income activity can be observed in AN hamlet's neighboring villages. When the author HWJ visited Xepon District in September 2022, many households in a neighboring village of AN hamlet were participating in hired labor work on rubber plantations. These households still practiced swidden agriculture, but as they spend more time working on rubber plantations, they find it more difficult to devote time to swidden agriculture. Thus, they were forced to hire workers from outside the household to keep the swidden agriculture operations going.

The increased number of workers contributed to the increased total work hour resources of AN hamlet. As aforementioned, although the sudden migration strained AN hamlet's land resources, the increased adult population contributed to the expansion of the village's work hour resources in the short term. However, the increased population is likely to be a labor resource to supply the rapidly developing plantations in Xepon District. As a result, the structure of subsistence in AN hamlet may undergo a major transition.

5.3 Reflection on the Classical Theory of Land Use Intensification

Boserup's (1965) theory of land use intensification focused on the input of land and time resources to the "main subsistence" and its contribution to survival. However, the case of AN hamlet was difficult to explain with Boserup's theory. As mentioned above, in the case of AN hamlet, residents have spent as much time on their subsidiary subsistence as on their main subsistence. From the results of the time allocation survey, we can infer that subsidiary subsistence are extremely important as a survival strategy for the residents of AN hamlet. To explain why this came to be, it seems that the development of a move toward a market-oriented economy plays an important role.

Most of the materials Boserup used in her analysis came from the surveys conducted in the communities in 1960s and 1970s, or earlier, which had even less marketization than the present-day AN hamlet. In the past, products gained from hunting and gathering, fishing, and other subsidiary subsistence were rarely used except for self-consumption or as gifts within the community. According to interviews with residents of AN hamlet, up until the 2000s, they did not sell hunting and gathering products to the market. During the days when motorbikes were not available, the Xepon District market 20 km away was a distant presence. However, if they catch a somewhat large prey (for example, bamboo rats) now, they can cross the river immediately and go to sell them in the Xepon market. Although the river is an obstacle, middlemen come up to the opposite shore and buy forest products such as bamboo shoots. The commodification of natural resources through these subsidiary subsistence contributes to a greater or lesser degree to the survival of AN residents and alleviated the problems such as land resource crunch brought about by the high-paced development occurring nearby.

The change to a market economy may also be affecting the positioning of all types of subsistence and the carrying capacity of the village ecosystem. When investigating land use intensification, the market factor should also be incorporated. Furthermore, to understand the livelihood of villages that are in the process of becoming a market economy, such as AN hamlet, it is necessary to observe not only traditional main subsistence, but also the entire subsistence activities including subsidiary subsistence.

6 Conclusion

The authors used the results of a time allocation survey to analyze the actual situation of time allocation in rapidly developing AN hamlet from the perspective of time geography. Considering classical land intensification theories, AN hamlet is already in a situation wherein its subsistence structure must change. However, the commodification of products obtained through subsidiary subsistence activities is alleviating this situation. In this sense, unlike the era in which classical land consolidation theories emerged, when considering land use in local societies where market economies are becoming more widespread, it is necessary to be mindful of the contributions of not only the main subsistence but also subsidiary subsistence to carrying capacity. These market economies are also bringing in diverse new ideas to these local societies, such as cash crops and employed labor, creating the possibility of various changes. Moving forward, time allocation surveys will likely be an important method for understanding residents' livelihood changes in local societies.

The spread of the market economy, pressure on land resources, limited family labor time resources, increased total village labor time resources due to immigration, and the flow of labor time to higher income subsistence activities are some of the many factors that are accelerating subsistence change among AN residents. Future research must continually observe how swidden agriculturist in AN hamlet and central Laos accept these changes and how they go about adjusting their lives, in terms of not only time allocation but also population reproduction and health. Another important challenge going forward will be to collect and analyze data on changes in the balance between cultivating swidden and letting swidden lie fallow, including the processes involved therein. Examining these data on shifting field as the basis for resident production will facilitate the understanding of the factors contributing to the intensification of land and time resource use in swidden agricultural society in the process of becoming a market economy.

Places such as AN hamlet exist in the frontier regions of Laos and across the world. The issue of how to improve the residents' lives and advance beneficial development while maintaining their living environment is an important challenge not only for residents but also for regional governments, development enterprises, and researchers.

In this study, we presented a time allocation survey method using recall, GPS loggers, and accelerometers, and demonstrated its usefulness based on the data collected. We then demonstrated the feasibility of conducting time allocation surveys in communities in developing countries using spatial statistical analysis methods and land use survey methods. By adding individual socioeconomic attributes to these survey methods and analytical techniques, we can expect to reveal characteristics of spatiotemporal changes in human activity.

In this study, as Table 1 shows, we conducted time allocation survey a total of three times, on 25, 13, and 49 participants, respectively. At the end of daily survey during the study period, the status of GPS loggers and accelerometers were checked, and at the same time one surveyor conducted a short interview about the daily activity detail using a recall coding sheet, without significant burden to the participants. If the number of surveyors were increased, it would be possible to collect activity data of even more subjects.

To the authors' knowledge, there have been no reports evaluating the spatial distribution of physical activity intensity (energy expenditure) using the records of accelerometer and GPS logger as in this study. In this study, we merged the two types of data from GPS logger and accelerometer using time stamps recorded by these two instruments. Using these data, it was possible to show the gender differences and seasonal changes in the spatial distribution of the survey participants' physical activity intensity, as shown in Fig. 5. In the future, we expect this will be used to collate activity content using chronological physical activity intensity and location information, as well as be applied to assessments of labor intensity.

However, some efforts are necessary to ensure data quality. This implies recruiting a sufficient number of informants and creating a coding sheet by conducting a detailed survey, after establishing trust relationships with study village residents. Given that slashing is the main subsistence activity in the dry season, it can be inferred that the large change in active space area for adult men and women between February 2017 and 2020 shown by Fig. 3 is the result of the latter's increase in the number of informants, which contributed to the increase in active space area. Ensuring an adequate number of participants (e.g. 30 men and 30 women each) is therefore also key to the success of the survey. The development of activity coding tables based on careful interviews and observations is an essential element to ensure data quality. Regarding making land-use maps, which is important for survey and data analysis in rural societies in developing countries, the authors worked with landowners to trace the boundaries of shifting fields using GPS and produced land-use maps in two to three days. With this method, we obtained information on land use in a relatively short time and confirmed the status of swidden cultivation and vegetation. It was then made to possible to assess the relationship between daily life time allocation and the spatial distribution of shifting fields, which are the production base of the survey participants' main subsistence.

As mobile communications spread in rural parts of developing countries in the future, collecting information on human activity even without GPS logger could be possible. That said, while collecting large amounts of data with these methods could be possible, unless we understand the livelihoods of those survey participants, obtaining significant results would likely be challenging. The present authors believe that doing fieldwork with a two-prong approach using face-to-face interviews and a variety of tools will lead to a deeper understanding of the participants' lives.

The time allocation survey data presented in this study have several deficiencies. The first pertains to the issue of the length of the survey period. Time allocation surveys should normally be conducted in terms of one cycle within the daily lives (for example, one week) of the participants; however, we had to shorten the survey period owing to weather and village-wide events at the time. The second issue is the small number of survey participants in the first and second surveys. This issue had the same cause as the first one. Devising solutions for recruiting participants is essential for ensuring data quality. In future surveys, the authors consider that this situation should be taken into account to allow for a longer reserve survey time.

Acknowledgements This study was supported from JSPS Grants-in-Aid JP25257004 and JP17H01633. We would like to thank Lao Tropical and Public Health Institute and the Savannakhet Provincial Health Department for their cooperation during fieldwork. We would like to thank Dr. Sengchanh Kounnavong, Dr. Somphou Sayasone, and Dr. Tiengkham Pongvongsa for their advice on our survey plan. We would like to thank Dr. Inthanomchanh Vongphoumy, Dr. Phet Nansounthavong, Mr. Monethong Inthisith, Mr. Lamsee Phanthakeo, and Mr. Hongsa Sinpaseuth for their kind help during the field surveys. We would also like to thank the people of AN hamlet, who kindly participated in the survey each day without fail.

References

Boccalandro F (2010) Exercise stress testing. In: Levine GN (ed) Cardiology secrets, 3rd edn. Mosby, Maryland Heights, pp 54–59

Boserup E (1965) The conditions of agricultural growth: the economics of agrarian change under population pressure. George Allen & Unwin Ltd., London

Carlstein T (1982) Time resources, society and ecology: on the capacity for human interaction in space and time, vol 1. Preindustrial society. George Allen & Unwin Ltd., London

Castree N, Kitchin R, Rogers A (2013) Time-geography. In: Rogers A, Castree N, Kitchin R (eds) A dictionary of human geography. Oxford University Press, Oxford, pp 516–517

Chayanov AV (1966) The theory of peasant economy. For the American economic association translation series. Originally published in Russian in 1925. Richard D. Irwin Inc., Homewood

Hirsch JA, Winters M, Clarke P, Mckay H (2014) Generating GPS activity spaces that shed light upon the mobility habits of older adults: a descriptive analysis. Int J Health Geogr 13:1–14. https://doi.org/10.1186/1476-072X-13-51

Jiang H, Lin L, Yonto DA, Pongvongsa T, Kounnavong S, Moji K (2021) Association between physical activity and activity space in different farming seasons among rural Lao PDR residents. Trop Med Health 49(73):1–10. https://doi.org/10.1186/s41182-021-00364-6

Jiang H, Sato R, Yokoyama S, Nishimoto F (2023) Temporal-spatial distribution of swidden agriculturists' daily activities in central Laos. E-J GEO 18:324–338 (in Japanese). https://doi.org/10.4157/ejgeo.18.324

Kumahara H, Schutz Y, Ayabe M, Yoshioka M, Yoshitake Y, Shindo M, Ishii K, Tanaka H (2004) The use of uniaxial accelerometry for the assessment of physical-activity-related energy expenditure: a validation study against whole-body indirect calorimetry. Br J Nutr 91:235–243. https://doi.org/10.1079/bjn20031033

Kwan MP, Lee J (2004) Geovisualization of human activity patterns using 3D GIS: a time-geographic approach. In: Goodchild MF, Janelle DG (eds) Spatially integrated social sciences. Oxford University Press, Oxford, pp 48–66. https://doi.org/10.1093/oso/9780195152708.003.0003

Minge-Klevana W (1980) Does labor time decrease with industrialization? A survey of time-allocation studies. Curr Anthol 21:279–298. https://doi.org/10.1086/202455

Moji K (1985) A time-saving spot-check method applied to a Sundanese peasant community in West Java. Man Cult Ocean 1:121–127

Nishimura Y, Okamoto K, Boulidam S (2010) Time-geographic analysis on natural resource use in a village of the Vientiane Plain. Tōnan Ajia Kenkyū (Jpa J Southeast Asian Stud) 47:426–450. https://doi.org/10.20495/tak.47.4_426

Noma H (1988) The context of time allocation studies in peasant societies: a review of developing countries' case. Jinbun Chiri (Jpn J Hum Geogr) 40:144–163 (in Japanese)

Ohtsuka R (2011) Ningen katsudou to seigyou tekiou (Human activity and subsistence adaption). In: Watanabe C, Umezaki M, Nakazawa M, Ohtsuka R, Sekiyama M, Yoshinaga J, Moji K (eds) Ningen no seitaigaku (Human ecology). Asakurashoten, Tokyo, pp 69–81 (in Japanese)

Patterson Z, Farber S (2015) Potential path areas and activity spaces in application: a review. Transp Rev 35:679–700. https://doi.org/10.1080/01441647.2015.1042944

Phouyyavong K, Tomita S, Yokoyama S (2020) Smallholder's labor allocation for livelihood diversification: a case study in an upland village in northern Laos. Tropics 29(1):9–24. https://doi.org/10.3759/tropics.MS19-08

Sato R, Jiang H, Nishimoto F, Yokoyama S (2025) Food procurement strategy of swidden agriculturalists in central Laos. In: Yokoyama S, Niwa T, Jiang H (eds), Population dynamics and livelihood changes of small-scale societies in Laos. Springer, Singapore, pp 19–43. https://doi.org/10.1007/978-981-96-2956-5_2

Thrift NJ (1977) An introduction to time-geography. Geo Abstracts Ltd., University of East Anglia, Norwich

Yamauchi T (2018) Evaluation of everyday life: time-space allocation and physical activity. Nihon Shoku Seikatsu Gakkai Shi (J Integr Study Diet Habits) 28(4):247–252 (in Japanese). https://doi.org/10.2740/jisdh.28.4_247

Population Dynamics and Livelihoods in a Paddy Farming Community in the Plains

Population Growth and Generational Response to Livelihoods in a Rain-fed Paddy Farming Village of Central Laos

Shinichi Takahashi, Satoshi Yokoyama, Futoshi Nishimoto, and Takahito Niwa

Abstract In a rain-fed rice farming village located in central Laos, we clarified the relationship between long-term population growth exceeding 2% per year and changes in livelihoods and various activities of villagers by generation. From the first to the fourth generation in the village that was established in the late 1920s, the acquisition of the paddy fields, which was the key to livelihoods, changed from opening paddies to inheritance and then to purchase along with the population growth. Migration to other rural villages increased from the third generation who did not acquire sufficient paddy fields due to population growth. In the fourth generation who experienced continued population growth, the shortage of paddy fields was alleviated by the emergence of circular migrant workers to Thailand. The responses to population growth in this village did not improve agricultural productivity or migration to urban areas instead of the acceleration of rural-to-rural migration against the background of the expansion of paddy fields in central Laos and the increase in number of circular migrant workers to Thailand. In recent years, the spread of family planning has contributed to the end of population growth, which will likely change the direction of rural livelihood transition in this region.

Keywords Laos · Population growth · Generation · Paddy land acquisition · Rural-to-rural migration · Circular migration to Thailand

---

This chapter was a revised version of Takahashi et al. (2023).

S. Takahashi (✉)
Kobe University, Kobe, Japan
e-mail: takahashi@econ.kobe-u.ac.jp

S. Yokoyama
Graduate School of Environment Studies, Nagoya University, Nagoya, Japan
e-mail: yokoyama.satoshi.i4@f.mail.nagoya-u.ac.jp

F. Nishimoto
Embassy of Japan in Laos, Vientiane, Laos

T. Niwa
College of Human and Social Sciences, Kanazawa University, Kanazawa, Japan
e-mail: niwa-2@staff.kanazawa-u.ac.jp

S. Yokoyama et al. (eds.), *Population Dynamics and Livelihood Changes of Small-Scale Societies in Laos*, International Perspectives in Geography 22,
https://doi.org/10.1007/978-981-96-2956-5_4

1 Introduction

Least developed countries have not been particularly involved in the market economy due to their historical political circumstances, with livelihoods in many such countries depending on subsistence agriculture. Therefore, these countries have had the lowest levels of per capita GDP worldwide. However, even these countries have experienced periods of population growth at 2–3% per annum for over 40 years, which was also facilitated by a lower mortality rate after World War II (United Nations 2022). This meant a doubling of population within a single generation of 30 years, a rate rarely seen in most developed countries. Subsequently, the birth rates in many least developed countries started declining from the 1990s, albeit behind so-called emerging nations and others, and the population growth has recently slowed. In addition, despite the continuation of significant population growths in many of these countries, lagging behind other developing countries, experienced comparatively high levels of economic growth mainly since the 1990s because of varied reasons (World Bank 2023).

How did people in places where subsistence agriculture is the livelihood of a large proportion of the population respond to this striking and long-term population growth? Did this result in an increase in the number of poor people due to an imbalance between the population growth and provisions need to survive (Malthus 1798), did the population growth consolidate agriculture, thereby creating a cornerstone of economic development (Boserup 1965), or was there a completely different path?

This study focuses on how livelihoods respond to this type of rapid population growth. However, in reality, this type of problem setting is not easy. In particular, changes in livelihoods are influenced not only by population growth but also by the economic, social, and political conditions of the time. In addition, regional differences arise in changes in livelihoods in response to population growth, depending on surrounding environment, such as natural conditions, types of agriculture, relationship with urban areas, population density, etc. When looking at subsistence small-scale collective societies from this analytical perspective, the most basic units that are directly involved in population growth and other factors that bring about changes in livelihoods are rural households or families that are responsible for production and population reproduction (de Sherbinin et al. 2008: 39).

This chapter is a monograph that clarifies the response mechanism of livelihoods to a rapid population growth through an analysis of the actual situation since 1920s, using data from mainly households or families interview surveys of one rain-fed paddy farming village in Laos, a least developed country located on the eastern side of mainland Southeast Asia. Laos has experienced upheaval, including colonization by France at the end of the nineteenth century, independence in 1953, the subsequent civil war, and the establishment of the Lao Socialist State in 1975. However, it is thought that the rural areas have hardly changed for long time since colonization (Stuart-Fox 1997). Although there are differences between swidden agriculture in the mountainous areas and rain-fed agriculture in the lowlands, subsistence agriculture dependent on the weather has been the norm. Small-scale farmers in each village have been the unit of livelihood and life. The civil war caused severe damage to ethnic minorities in the mountainous areas and some lowland farmers, forcing them to evacuate or move. Even so, their basic subsistence livelihood has remained almost unchanged.

This lack of apparent change can also be seen from another perspective, as the farmers' behavior was independent from outside political and economic influences. In actual rural Laos, there always seems to have been a dynamic change, such as the response of livelihoods to population growth. It is possible to show how vibrant rural areas of present-day Laos were in the pre-modern period (Rigg 2005). It was only after the 1990s that the transition to a market economy and globalization became more pronounced that rural Laos began to experience different drastic and unavoidable change and send out many circular migrant workers to big cities like Vientiane, Laos and Bangkok, Thailand.

Thus, we think rural villages in Laos, which have undergone various changes while at the same time maintaining their livelihoods through subsistence rain-fed agriculture, are appropriate for clarifying the relationship between population growth and livelihood.

2 Analytical Framework and Data

2.1 Population Growth and Response by Generation

There has been considerable research to date based on a fact-finding investigation on how livelihoods of farming villages in many countries have responded to population growth or population pressures (For example, Bilsborrow 1987; de Sherbinin et al. 2008; Carr 2009; Jane et al. 2014). In particular, those studies were conducted in countries and regions, such as Asia, including Laos, Africa, and Latin America. Many such studies have analyzed the relationship between the population growth and various factors of livelihoods from either a macro- or micro-perspective. In general, many of these analyses take population growth as a given condition, or take a cross-sectional approach to the relationship between population growth and household livelihoods. Such research has obviously been valid depending on their research objectives (For example, Bongaarts 1996; Mwesigye et al. 2017).

Among these studies, in Japan, a study on Don Daeng Village in Northeast Thailand (see Fig. 1) by the Center for Southeast Asian Studies, Kyoto University, conducted a fact-finding investigation of population growth and change in livelihoods at the micro level, which has produced many results (For example, Mizuno 1981; Kuchiba 1990; Fukui 1993). Regarding the relationship between the population growth and livelihoods of villagers and other economic activities, first feature of the study for Don Daeng Village is the in-depth analysis of the population growth and the change in its factors such as birth, death, and migration. An important perspective is the interrelationship between people's reproduction, such as birth, death, and marriage, which affects population growth, and people's production, which relates to livelihood activities. A second feature is that the research on Don Daeng Village considers the long-term population growth over a period of 80 years and the relationship with household livelihoods, such as rice cultivation, paddy fields, migration, and nonagricultural employment. When looking at the relationship with population growth, the population growth changes over time and the factors causing such a change (i.e., births, deaths, and migration) are always changing. Such changes also impact the changes in household livelihoods.

This chapter focuses on four analytical methods while taking into account the features of the Don Daeng Village research described above. First, we analyze the relationship between population change considering the factors of births, deaths, and migration, and the change in household livelihoods. Second, the change in household livelihoods over a long-term period of multiple decades is closely related to the acquisition of paddy fields and children's choice to migrate. Therefore, we incorporate a new analytical method that is not traditionally used by examining the genealogical relationships in generations between parents and children to clarify the population growth and the change in household livelihoods by generation. Third, regarding livelihoods, we not only consider the simple area of paddy fields but also incorporate a new analytical method to look at the relationship between population growth and paddy field acquisition data, categorizing the acquisition of paddy fields by households in each generation by reclamation, inheritance, purchase and no acquisition of paddy fields. Fourth, the above-mentioned relationship between population growth and livelihoods can carry an imbalance in people's lives, and there are various responses

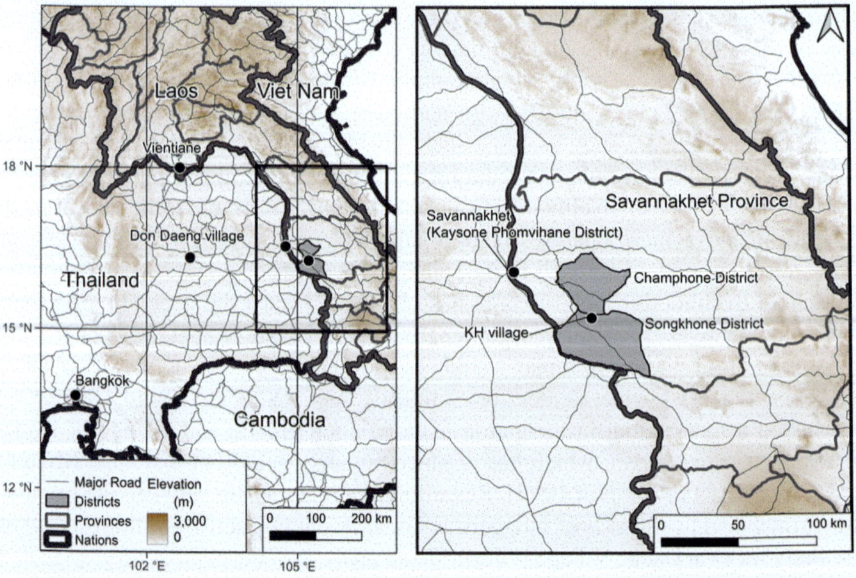

Fig. 1 Location of KH village

to the imbalance. Referring to Davis' "multiphasic response theory" (Davis 1963), we can provide specific conceivable responses for farming village regions such as Laos as follows:

(1) Response from agriculture

Expansion of farmland through reclamation; intensive use of farmland as Boserupian response (Boserup 1965); purchase of farmland from nearby farming villages.

(2) Introduction of nonagricultural sectors within the region

Small retail or tourism businesses; introduction of rural industries, such as textiles.

(3) Absorption of the population growth

Commuting to work in nearby urban areas; circular migration to domestic urban areas; and circular migration overseas.

(4) Suppression of the population growth

Permanent migration to rural or urban areas; overseas permanent migration; birth control.

(5) Equilibrium between livelihoods and population growths, whereby the population is affected by an increase in the mortality rate caused by so-called impoverishment in response to the sharing of available resources as Malthusian response (Multhus 1798).

The analysis in this chapter shows these response choices and factors as explicitly as possible.

The analytical outline of the chapter is as follows. We look at the types of responses in the acquisition of paddy fields, focusing on their reclamation, inheritance, and purchase, which is the main livelihood accompanying the significant population growth for the four generations since the founding of the village in the 1920s up until the present. In addition, in the case of no acquisition of paddy fields, we clarify the role of behavior, including the migration of people to other regions, particularly the rural-to-rural permanent migration and the recent circular international migration to Thailand, through the results of the field survey of the change in the livelihoods of each generation. From this point on, we would like to highlight the unique response to the population growth in a Laotian farming village in a low population density region that is typical of many regions across Southeast Asia. The survey area is KH village, Songkhone District, Savannakhet Province, Central Laos (Fig. 1). This village is engaged in subsistence rain-fed paddy agriculture, a typical form of agriculture in Laos. This has been a common form of agriculture also in Thailand and other Southeast Asian countries.

2.2 Analytical Data

The data used in this chapter were obtained through following four surveys, conducted in KH Village and Bangkok from 2013 to 2015. The first is based on two genealogical surveys conducted in September and December 2014. The genealogical surveys collected information about every individual resident's sibling relationship, their parents and their siblings, and their grandparents and their siblings, then took the name, place of birth, spouse, and the age (calendar year of birth), and ages or calendar years of death, migration, and marriage for each person, thereby grasping the locational relationship of each person's genealogy to produce a family tree. Second is complete surveys of matters such as the population of households, migration of people, agricultural production, circular migration, and family planning conducted in December 2013, December 2014, and November 2015. The third is a survey concerning the acquisition of paddy fields, using high-resolution satellite data to measure the area of paddy fields, and interview surveys on reclamation, inheritance, and purchase of paddy fields since the village was founded. The fourth is interview surveys to circular migrants to Thailand in the village and surveyed circular migrants from the village residing in Bangkok in 2015.[1]

3 Overview of the Study Area and Population Growth

3.1 Study Area

The study area KH village is located approximately 70 km southeast of the provincial capital of Savannakhet (officially named Kaysone Phomvihane District) (Fig. 1). The village and surrounding rural areas are spread out on the plain 130–150 m above sea level, where a tributary of the Banghiang River flows into the Mekong River, making it possible to develop paddy fields. The main livelihood of these farming villages is rain-fed paddy rice cultivation. The 2014 de jure population[2] of the village was 755, with a circular migration to Thailand population of 260 (34% of the de jure population), so the de facto population was 495 (66% of the de jure population). There were 110 households. Laos has a tropical monsoon climate separated into the dry season (November to April) and wet season (May–October). The average annual temperature and rainfall from 2005 to 2014 in Savannakhet Province, to which this

---

[1] Part of the results presented in Niwa et al. (2021).

[2] De jure population is the so-called population with their place of residence in the village and includes people who temporarily left the village for circular migration or other reasons. Circular migration involves temporarily moving away from the village, even if one does not return for over one year, unless the place of residence does not change due to marriage, etc., treated as de jure population in the village. de facto population consists of all persons who are present and living in the village. In the case of this village, the population can be considered as de jure population minus circular migrating population.

village belongs, were 26.0 °C and 1,535 mm, respectively (Lao PDR, Statistics Bureau 2015c).

Rice cultivation in rain-fed paddies is undertaken in the wet season. According to water availability due to different altitudes in the microtopography, paddy fields can be divided into low-lying rice fields that are susceptible to floods, mid-lying rice fields, and high-lying rice fields subject to the impact of drought. Low-lying rice fields are the most productive. However, changes in production volumes can be drastic depending on frequent floods and droughts, and the village alternates between good harvests of 3 t of unhulled rice per ha and bad harvests of less than 1 t per ha. Therefore, the average annual land productivity of the village's rain-fed paddies is approximately 1.5–2 t of unhulled rice per ha. Given such unstable conditions, there is a tendency to be cautious about investing in ways to boost land productivity in this village. Regarding fertilizers, organic fertilizers obtained by allowing livestock to graze in the paddy fields and chemical fertilizers have been used for a long time. Regarding irrigation, except the recently started 3 ha of irrigated rice cultivation, irrigation has not been used.

Other than rice cultivation, the villagers produce vegetables for their own consumption, feed livestock and collect nontimber forest products (NTFPs). Livestock comprise water buffalo, which previously had a central role in plowing paddy fields, cattle, and pigs, as well as poultry, such as chickens and ducks. Water buffalo and cattle are mainly personal property and bred to generate funds for agricultural machinery, house construction, education, hospital admissions, and other requirements. Taking advantage of the natural conditions, they are kept in the paddy fields in the dry season and the village woodlands in the wet season. The villagers currently purchase a high proportion of pig and poultry feed. Forest collection includes plants, such as bamboo shoots, and animals, such as squirrels, insects and so on. Unlike swidden regions, there is little common forest area, so a high proportion of NTFPs is for self-consumption. Although some NTFPs are sold in the village or to merchants from outside the village, this accounts for a small proportion of gross income. Overall, a large income comes from livestock sales, and some farmers sell rice in years with good rice harvest. Although there is diverse agricultural production, rice cultivation undeniably remains the most important livelihood for the villagers.

On the other hand, in terms of nonagricultural employment within the village, the most important has been circular migration to Thailand, particularly Bangkok and its surrounding areas. Circular migration to Thailand has become pronounced since the 1990s, and now, almost all young people under the age of 30 years are engaged in circular migration (Niwa et al. 2021: 21–22). The only commuter residing in the village is a teacher who travel to an elementary school in a nearby village, and there are no nearby urban areas that can absorb employment. There are several carpenters, and five households operate small general stores; however, all of these are side businesses to agriculture.

Taking a look at the beginning of the village, the first settlers that built the current KH village are said to have come from other places in the late 1920s. The initial population is estimated to have been about 60. Since its founding, including those who have already died or emigrated, the population comprises a family tree of 1,668 persons. Among these, there are 37 persons whose genealogy could not be traced further up the line. Some of them are clearly stated that they had moved from a nearby village.

The 37 persons in the first generation were 19 males and 18 females, and they comprised 18 married couples.[3] All the 110 current households in KH village can trace their roots back to these 18 married couples. Because of a range of the estimated dates of birth, it is reasonable to assume that the first settlers did not arrive at the same time but arrived intermittently between the 1920s and 1940s. In addition, despite immigrating, it is conceivable that some did not leave descendants or migrated again to different regions and that such cases are not remembered by current residents. Several children (second generation) of the first settlers were still alive at the time of the survey. In addition, the largest population is currently the fourth generation, with the latest generation being the fifth generation. Assuming it takes 25–30 years for generational change, it took approximately 80 years for the fourth generation to appear. The nearby villages from where the first settlers were sent to KH village are old villages and have histories of nearly 300 years, so by comparison, this village can be called a new village. KH village is an agricultural frontier that was late to development, and it is conceivable that the primary reason for forming this village was to absorb the population growth in nearby villages.

3.2 *Population Change in KH Village*

Before looking at the village's population, we provide a general overview of the population change in Laos as a whole. Laos conducted its first population census in 1985, and at the time, its population was 3.58 million (Lao PDR, Statistics Bureau 2015a: 21–22). There were no previous demographic surveys on matters such as births and deaths, with this census being the first that also surveyed births and deaths. Population estimates before the census was announced by its former colonial power France, and the government of Laos after independence in 1953, but they are said not to be accurate reporting of population (Halpern 1961a: iv). Therefore, we must look at the changes in the population prospects from the United Nations (United Nations 2022) since World War II. Although this is also not accurate, we can get a rough idea of the trend. The population prospects state that Laos' population was 1.68 million in 1950, 2.68 million in 1970, 3.25 million in 1980, 5.32 million in 2000, and 6.6 million in 2015. The annual population growth rates from 1950 to 2000 were close to 2.5%, despite there being a temporary decline due to a civil war in the 1970s. Thus, there were striking increases in population. The average annual growth rate since 2000 has dropped below 1.5%, thereby moderating the population growth.

Next, we look at the population growth in KH village. Data on population, births, deaths, and migration are obtained by tabulating genealogical survey results. As shown in Fig. 2, there has been a rapid population growth since around the 1920s

[3] One woman divorced remarried another man who moved to the village.

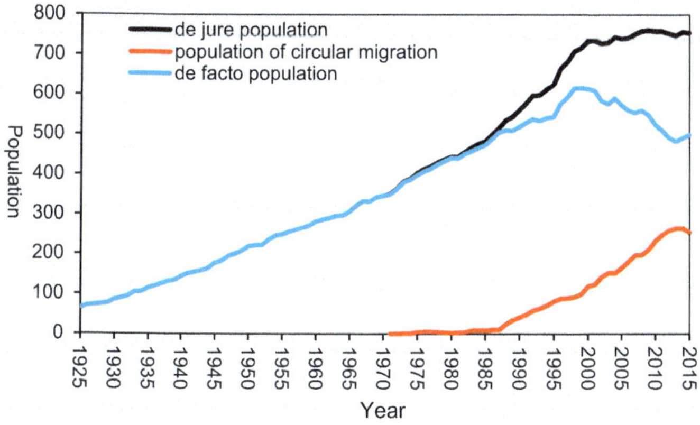

Fig. 2 Population changes in KH village, 1925–2015

when the village was established. This village and the surrounding rural areas were not heavily damaged by the civil war of the 1960s and early 1970s, and the population did not decline but continued to grow. Though the growth rate was lower from 2000 onward, this has been the annual growth rate of 3% for approximately 90 years since the founding of the village, and the population exceeded 750 in 2014. This population is the de jure population, that is, permanent resident population that base themselves in the village. As mentioned above, temporary circular migrants to Thailand are included in the de jure population. There has been a dramatic increase in the population undertaking circular migration to Thailand since the 1990s, presently accounting for over 35% of the de jure population. Therefore, the de facto population of people actually residing in the village has reduced as the circular migration to Thailand has increased since the late 1990s, reaching about 500 in 2014.

We look at this population change by decomposing into (crude) birth and (crude) death rate, and permanent in-migration and out-migration rate (Fig. 3), where population growth rate = birth rate − death rate + in-migration rate − out-migration rate. In the 1920s–1930s when the village was founded, livelihoods were clearly focused on acquiring new land for paddy fields for rice cultivation. At the time, the influx of young people and the number of their children (including many births) created a rapid population growth. The average annual growth rate in the 1920s–1940s reached 4–6%. Since the 1950s when land reclamation stabilized, the birth rate was in the 40s per 1,000 population. This high level continued until the early 1990s through natural fertility[4] without conscious birth control. The birth rate then declined sharply to around 15 per 1,000 population by 2010. Dissemination activities for family planning in the village commenced in the late 1990s, and the birth

[4] Married women before 1970 had no experience of contraception.

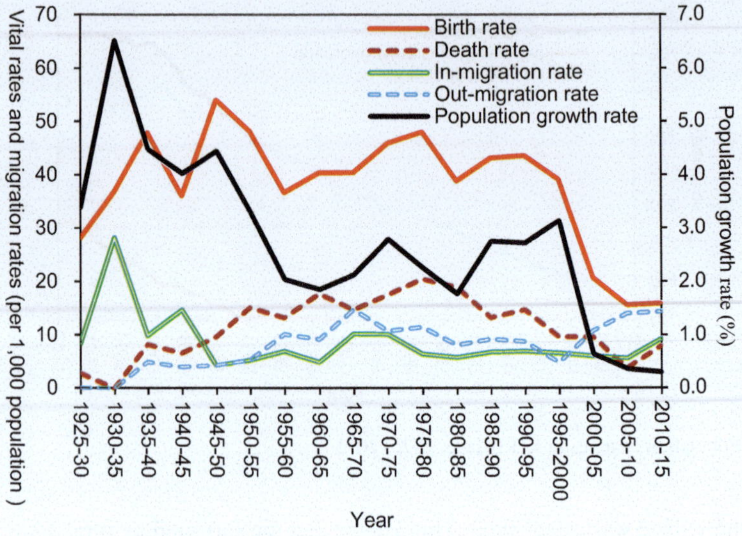

Fig. 3 Changes of population growth rates, vital rates, and migration rates, 1925–2015

rate declined because of an increase in the number of women taking contraception. Important factors contributing to a high birth rate were early marriages[5] and universal marriage, that is, almost everyone marrying. These trends of early and universal marriage remained even though there has been a slight increase in the unmarried rate due to the recent increase in circular migration.

The death rate was low in the 1930s–1940s because of the high proportion of young people and the difficulties in recording infant deaths at that time. Subsequently, by the late 1970s, the death rate was roughly 15–20 per 1,000 population, but there was a slight upward trend. This change is attributed to several factors, including improvements in the recording infant deaths,[6] the change in the age structure, and the impact of the civil war. In addition, in the 1960s, a hospital was constructed in the central area of the district in which the village is located. However, due to the civil war at the time, the hospital did not function adequately. There was a pronounced decline in the death rate from the late 1980s. With the establishment of the revolutionary government in 1975 and the introduction of the market economy in the late 1980s, the increase in development aid from international institutions, such as the WHO, and foreign governments advanced the spread of modern healthcare. Furthermore, such stimuli led to active medical care through the government maintenance of the district hospital and the spread of public health to villages (Nakamura and Iwasa 2008). Consequently, the death rate recently fell to below 10 per 1,000 population.

---

[5] The average age at the time of the first marriage for married women from 1990 to 1999 was 20.0 years and 20.6 years since 2010.

[6] The newer the era, the greater the ability to directly obtain more accurate information from living people.

The permanent migration in this village is characterized mostly by marriage. About 80% of the migrants are aged below 30, with a trend for concentration in the young. The rates of migrants from outside the village were temporarily exceeding 20 per 1,000 population during the reclamation period of the 1920s–1930s, but declined from the 1940s. Subsequently, despite a slight increase since the 1970s, it remains below 10 per 1,000 population. Nevertheless, this in-migration rate increased slightly from 2010 onward, rising to 10 per 1,000 population. This is, as noted later, mainly due to the increase in young people remaining in the village but engaging in circular migration and the increase in spouses arriving from outside the village. Conversely, the out-migrant rate gradually increased after reclamation and has been almost continuously in excess of 10 per 1,000 population since the 1950s, thereby exceeding the level of in-migration rate. The out-migration rate has further increased since 2000, which can be attributed to the outflow of marriages related to the circular migration to Thailand.

Lastly, we look at the fertility of married women (completed fertility) by birth cohort and the infant mortality rate in the village. According to the results of the village survey, the average number of births for the female birth cohort up until the 1960s, when there was no birth control, was 5.7. Given the likelihood of incomplete grasp in past results in the survey, the actual figure may be slightly higher.[7] For the married women birth cohort since the late 1960s, the average number of births suddenly fell due to advancements in provincial and district family planning from the late 1990s. For the 1970s birth cohort who are close to their period of completed fertility (ages of 35–44 years as of 2015) the number of births was 3.2, indicating fertility level through birth control.

The infant mortality rate for the married women cohort born before the 1970s was 165 per 1,000 births.[8] Thus, even if the married women cohort born before the 1970s bore 5.7 children, according to calculations using the model life table (Coale and Demeny 1983), only 4 survived infant children to reach adulthood. Infant deaths are small cases so the results have often random variety, but the figure fell to about 130 after the 1970s and fell sharply to an average of about 20 per 1,000 births after the 1980s due to the aforementioned maintenance of provincial and district hospitals and improvements in public health.

The results of the change in each of the factors of population growth outlined above led to high population growth rates for this village at least 2–3% per annum for a long time from the 1930s to 1990s. Taking one generation to be approximately 30 years and looking at this population growth in terms of units of married couples, the difference in the number of married couples between one generation and the

[7] The average completed fertility of married women born between 1900 and 1904 in county areas of Japan, that is, before the fertility decline in Japan was 5.2 (Statistics Bureau, Prime Minister's Office, Japan 1963).

[8] This is a high level, but it can be considered to be the same level as the infant mortality rate in Japan around 1900–1920. Japan's infant mortality rate at the time is likely to have had some omission in the survey, so the actual figure might be slightly higher (National Institute of Population and Social Security Research 2022).

next is likely more than double.[9] However, since then, the population growth has suddenly slowed due to a decline in the birth rate and increase in the out-migration rate. Recently, the net migration rate, which is the difference between the in-migration rate and out-migration rate, has become negative and the natural growth rate, which is the difference between the birth rate and death rate has declined. Consequently, there is sometimes a small but net decrease in the population.

4 Changes in Population and Livelihoods in Terms of Units of Married Couples

In this section, to directly determine the relationship between population growth and livelihoods, we identify the change by generation, particularly focusing on the acquisition of paddy fields which is the most important for the livelihoods of household units. Although households are essentially the units of livelihood, it is difficult to specify what constituted a household in the past, so we take the family of a married couple to be the unit. Laos is a multiethnic country, but in this village, all are ethnic Lao, who account for half of the population in Laos. In general, the ethnic Lao are said to have an inherited equal division of property like paddy fields. However, not everything is always equally divided, and it is often the case that the youngest child, particularly the youngest daughter, lives with and takes care of her aging parents while also inheriting a majority proportion of the paddy fields. Families generally comprise two-generation families, and there are very few three-generation families wherein married couple lives with their parents while the parents are working. However, it is common for children to continue living with their parents for a few years between marrying and building their own houses. In addition, as often seen in Northeast Thailand, children build houses near their parents after getting married in so-called multihousehold compounds to work and eat together with their parents (Mizuno 1981: 117–126), but there were few such examples in KH village. The general structure of families is that children become independent a few years after getting married, thereby maintaining a long-term nuclear family until being cared for by their children in their old age. Therefore, we believe there are few issues if we look at married couple families as the basic unit.

[9] In the case of population growth rates at 2.5% annually, it will take about 30 years for the population to double; thus, in a society such as Laos where almost every person who survives to adulthood marries, the married couples for one generation can be replaced with double the number of married couples in their children's generation in 30 years.

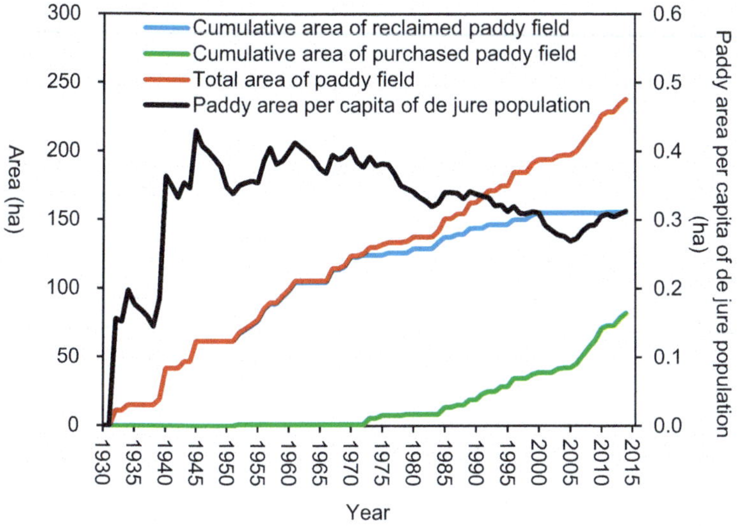

Fig. 4 Changes of areas of paddy reclamation and purchase and paddy area per capita of de jure population, 1930–2015

4.1 The Relationship Between Population Growth and the Acquisition of Paddy Fields in the Entire Village

Firstly, we look at the relationship between the over 2% of population growth and the increase in paddy fields for the entire village. Figure 4 shows the trend in the acquisition of paddy fields separated into reclamation and purchase since the village was founded, and the per capita paddy area for the de jure population. The area of paddy fields surged in the 1920s–1930s, when people who mainly arrived in the village from nearby villages created paddy fields by reclaiming the forest. Up until around 1940, the influx of villagers settled down and the growth of the paddy fields by area per capita ceased. From then until around 1970, reclamation continued, and there was an increase in population, but the paddy area per capita remained about 0.4 ha. Even if the production volume of unhulled paddy rice per area (ha) of the village at the time was lower than the present at an average of 1.5 t, unhulled paddy rice production per capita reached 600 kg by about 1970. Consequently, the amount harvested by the village was more than the 350 kg annual unhulled[10] paddy rice consumption per capita, so production was sufficient to support an increase in the village's population.

Since the 1970s, the proportion of reclamation has fallen with a decline in suitable places for paddy fields, particularly for the comparatively highly productive low-lying rice fields. Instead, the purchase of paddy field from siblings, relatives, and people in the village and nearby villages began in the 1970s, but initially the numbers were

[10] The following rice yields and consumption are all for unhulled paddy rice.

small. Therefore, the increase in paddy area has not kept pace with the population growth, and the paddy area per capita also declined. This was 0.3 ha per capita in the 1990s and fell further to 0.27 ha per capita in 2005. The volume of rice production in terms of the area per capita fell to 400 kg, close to the 350 kg annual paddy rice consumption per capita, which is enough to maintain the population but with little surplus. Since the 1990s, and especially since 2000s, there has been a significant increase in the purchase of paddy fields through remittances by circular migrants to Thailand, and the paddy area per capita exceeded 0.3 ha in 2015, so there was some surplus above the capacity to sustain the population.

The capacity to sustain the population that we have seen so far refers to the capacity per de jure population, and assuming that the population of circular migrants consumed little rice of the village, as discussed later, the cumulative paddy area per capita for the de facto population excluding circular migrating population from de jure population fell in the 1980s but reached 0.32 ha in 2005. The capacity to sustain the population is said to have risen with the progress in circular migration.

As noted above, when looking at the village as a whole, the paddy area per capita gradually fell as a result of a significant population growth that continued for a long time, but this did not reach dangerous levels; moreover, there was subsequently renewed surplus with capacity to maintain the population. We consider how this type of relationship between the population growth and change in paddy fields specifically manifested in the villager's lives and in paddy field agriculture that is their livelihood. In addition, we consider why the capacity to maintain the village did not drop to dangerous levels. We look at these issues through the change in livelihoods and population for each generation of married couple families.

4.2 Population Growth and the Livelihood Response by Generation: Changes in Paddy Field Acquisition and Migration

In this section, we trace the lineage of married couple families from the first generation that founded the village up until the fourth generation,[11] and analyze the relationship between changes in livelihood and population growths for married couple families. This method facilitates directly looking at the situation surrounding the livelihoods of parents and their children. However, given the differences among the village founders in terms of marrying age and timing of their move to the village, a single generation did not necessarily reside in the village at the same time. For example, the year of marriage for married couples that moved to found the village covers a wide span from 1911 to 1935. However, just under 80% of the married couples that moved to the village married between 1911 and 1929. The fourth generation includes married couples who married between 1974 and 2014, but more than 60% of these married couples married between 1990 and 2009. From the first to the

[11] Fifth generation is excluded because 70% of them are unmarried at the time of the survey.

Table 1 Paddy landholdings of married couples among village residents by generation

	No paddy field		Less than 1 ha	1–2	2–3	3–5	5–7	More than 7 ha	Total
	No inheritance	Circular migration							
1st generation	5	0	0	1	2	3	2	5	18
	(28)	(0)	(0)	(6)	(11)	(17)	(11)	(28)	(100)
2nd generation	10	0	4	5	6	13	5	3	46
	(22)	(0)	(9)	(11)	(13)	(28)	(11)	(7)	(100)
3rd generation	16	11	8	18	16	21	3	3	96
	(17)	(11)	(8)	(19)	(17)	(22)	(3)	(3)	(100)
4th generation	13	78	7	13	7	8	1	1	128
	(10)	(61)	(5)	(10)	(5)	(6)	(1)	(1)	(100)

Note The numbers in the table indicate the numbers of married couples, and the numbers in parentheses () indicate the percentages (%) of all married couples in the village by generation. The sum of the breakdown of percentages may not always add up to 100% due to rounding off

fourth generation, we take every generation to align with the years in which 60% or more of the married couples married. We do not believe this is a major issue when looking at the overall picture by generation.

In the following, we look at married couples in relation to the change in livelihoods, with a particular focus on paddy field acquisition and migration from the first to the fourth generation, during the rapid population growth.

4.2.1 First Generation (Primarily Married Before the 1930s)

Eighteen married couples in the first generation, who moved to the village in the 1920s–1930s, reclaimed 78 ha (4.6 ha per married couple) and not only secured rice but also prepared paddy fields for the next generation. Most married couples had five or more surviving children and nearly 60% of couples reclaimed more than 3 ha of land (Table 1). However, there were also five married couples without paddy fields in the initial period. These married couples are believed to have used their relatives' paddy fields who moved to the village at the same time.

4.2.2 Second Generation (Primarily Married in the 1940s–1960s)

In the second generation, reclamation gradually declined; only 20% of the married couples residing in the village had reclaimed forests and created paddy land (Table 2). Many of the married couples undertaking reclamation either did not inherit paddy fields from their parents or added reclaimed fields to their inheritance. The acquisition of paddy fields by a majority of this generation was mainly through inheritance. Conversely, married couples with absolutely no inheritance from their parents and without land or paddy fields accounted for 22% of the village residents. This also included elderly parents who continued to retain ownership, and ultimately, the land was inherited by their grandchildren generation. In addition, there may also have been cases wherein married couples did not own paddy fields but jointly cultivated those of relatives who had inherited paddy fields. Although we do not have detailed information about the time, land ownership was not as strict as it is now. It is thought joint cultivation, particularly among parents, children, and their siblings, has existed.

For the second generation, there was an ongoing increase in paddy fields through reclamation despite the main source being inheritance, and despite a natural increase through high fertility, the second generation was not subject to population pressures. For the entire village, the 46 married couples from the children of the first generation who survived and remained in the village had a total paddy area of 133 ha obtained through reclamation, inheritance, and purchase (only 1 married couple), amounting to 2.8 ha per married couple. Taking the average harvest at the time as 1.5 t per ha, lower than the current yield,[12] the paddy rice harvest per married couple was 4.2 t. Assuming that paddy rice consumption at the time was the same as the current level of 350 kg per adult in the village, the number of surviving children was 5, and the volume of their paddy rice consumption per capita was 70% of the adult level, the annual consumption for a married couple family was about 1.9 t. Even if the family comprised 7 adults, the amount would be 2.45 t, with a paddy area that could generate surplus rice.

Another feature of this generation is that 23 of the total 69 married couples left the village, which constitute not a small number of the total (Table 3). As shown in Table 4, many out-migrants in the 1940s–1950s, however including not a few unknown destinations, did not move to urban areas but migrated to nearby villages, farming villages within the province, or farming villages in other provinces. If all the married couples who left the village had stayed, the paddy area per married couple would have been about 2.0 ha. This area would have produced a harvest of 3.0 t, a rice harvest in excess of the 1.9 t level required for self-sufficiency. Thus, from an average perspective, 23 married couples left the village when there were surplus paddy fields. One conceivable factor was that in the second generation, there was little reclamation of low-lying rice fields with comparatively high productivity, so they moved to regions with ample uncultivated land that still had better conditions at the time.

[12] The Ministry of Agriculture in Laos conducted an agriculture survey in 1954, which reported the harvest of 1.34 t per 1 ha in KH village (Halpern 1961b: 7). This figure could refer to husked rice, but because it is unclear whether this was a normal year crop, we converted it by understating the unhulled paddy rice harvest at 1.5 t/ha (1.34 t of husked rice is converted to at least 1.8 t of unhulled rice).

Table 2 Paddy acquisitions and temporary migrants of married couples in the village residents by generation

	Reclamation	Reclamation and inheritance	Inheritance	Inheritance and purchase	Purchase	No paddy field		Total
						No inheritance	Circular migration	
1st generation	13	0	0	0	0	5	0	18
	(72)	(0)	(0)	(0)	(0)	(28)	(0)	(100)
2nd generation	9	6	17	3	1	10	0	46
	(20)	(13)	(37)	(7)	(2)	(22)	(0)	(100)
3rd generation	5	1	29	26	8	16	11	96
	(5)	(1)	(30)	(27)	(8)	(17)	(11)	(100)
4th generation	0	1	21	7	8	13	78	128
	(0)	(1)	(16)	(5)	(6)	(10)	(61)	(100)

Note is the same as in Table 1

Table 3 Permanent migrants and village residents among married couple by generation

	Village residents	Permanent migrants	Total
1st generation	18	0	18
	(100)	(0)	(100)
2nd generation	46	23	69
	(67)	(33)	(100)
3rd generation	96	105	201
	(48)	(52)	(100)
4th generation	128	120	248
	(52)	(48)	(100)

Note The numbers in the table indicate those of married couples, and the numbers in parentheses () indicate the percentages (%) of all married couples by generation

4.2.3 Third Generation (Primarily Married in the 1960s–1980s)

The turning point in the relationship to paddy field holding comes with the third generation. Despite the continued rapid population growth, this generation had virtually no suitable uncultivated land for reclamation, and although there were married couples that moved to the village and reclaimed land that did not have good conditions, reclamation effectively ceased around 1970.

As with the second generation, the number of married couples remaining in the village substantially increased for the third generation, thereby doubling from 46 in the second generation to 96 in the third generation. At the same time, there was diversification in the acquisition of paddy fields (Table 2). Married couples with reclamation only and reclamation and inheritance decreased. Couples with inheritance only, inheritance and purchase, purchase only, and no ownership of paddy fields increased. Among these, the proportion of inheritance only fell from about 37% in the second generation to 30% in the third generation. By contrast, there was an increase in married couples who acquired paddy fields through inheritance and purchase and purchase only, which together accounted for 35%, and the importance of purchasing paddy fields has increased. Paddy field purchases were mainly conducted by married couples who did not inherit paddy fields, married couples who inherited but supplemented their holdings due to a shortage of paddy fields, and married couples who owned paddy fields and wanted to expand. Additionally, their children (the fourth generation) engaged in circular migration, purchased paddy fields often in their parents' names with their remittances.

Compared to the second generation, the area of the paddy fields owned per couple decreased in this generation. Table 1 shows that the percentage of married couples with more than 3 ha of paddy fields was nearly 50% in the second generation, but dropped to 28% in the third generation. On the other hand, the actual number of married couples with paddy fields of less than 3 ha increased from 15 to 42. Those without paddy land also increased, and included agricultural laborers who were very few, and farmers who rented paddy fields and used paddy fields owned by their

Table 4 Permanent migration by gender, year, and region of origin and destination in KH village

Gender and migration mode	Migration period	Savannakhet Province		Other provinces	Urban areas	Thailand	U.S.A.	Unknown	Total
		Nearby villages	Other villages						
Male in-migration	1930–1949	11	0	0	0	2	0	4	17
	1950–1969	9	4	2	0	9	0	7	31
	1970–1989	22	5	10	2	0	0	4	43
	1990–1999	13	5	7	1	3	0	5	34
	2000–2009	17	5	3	2	2	0	5	34
	2010–2014	19	4	3	2	0	0	0	28
Male out-migration	1930–1949	2	0	0	0	0	0	5	7
	1950–1969	3	2	4	3	1	0	17	30
	1970–1989	18	1	4	6	7	1	12	49
	1990–1999	14	7	1	0	3	0	2	27
	2000–2009	24	9	1	1	5	0	10	50
	2010–2014	23	5	7	2	2	0	6	45
Female in-migration	1930–1949	9	0	0	0	1	0	7	17
	1950–1969	3	0	0	0	3	0	1	7
	1970–1989	9	3	5	0	0	0	1	18
	1990–1999	3	0	2	0	1	0	1	7
	2000–2009	3	2	3	0	0	0	1	9
	2010–2014	3	2	3	0	0	0	0	8
Female out-migration	1930–1949	1	0	0	0	0	0	1	2
	1950–1969	2	2	4	1	1	0	7	17
	1970–1989	14	5	1	2	5	3	3	33
	1990–1999	9	2	1	0	5	0	0	17
	2000–2009	12	3	4	1	10	1	2	33
	2010–2014	13	0	2	0	6	1	5	27

Note Neighboring villages are those in Songkhone and Champhone Districts (see Fig. 1)

children. Furthermore, some in this generation were engaged in circular migration to plantations in Central and Northeast Thailand in times of bad harvest due to floods or other factors. They were the so-called pioneers of circular migration. This type of circular migration developed and was the cornerstone of the subsequent significant increase in circular migration to Thailand, particularly Bangkok.

As with the second generation, we consider the relationship between the paddy fields and population growth per married couple. First, we assume the harvest per area, annual volume of paddy rice consumption, and ratio of child to adult consumption to be the same as the second generation. The average number of surviving children per third generation married couple was 4.1 which was lower than that of second generation married couples due to some married couples who married in the

1980s implementing family planning. In this case, the annual rice consumption per married couple family is taken to be about 1.7 t. In addition, the area of paddy fields possessed by third generation married couples was 206 ha with 96 married couples residing in the village. The average paddy area per married couple within the village was 2.1 ha, with paddy rice production of about 3 t. While this is adequate to maintain the families, there is not as much surplus as there was for the second generation.

Compared with married couples residing in the village and engaging in village livelihoods, the proportion of married couples who moved out from the village reached more than 50% of the total number of couples, which was 4.8 times the previous generation. Regarding permanent migration in the 1970s–1980s, the active period for the third generation, people mostly migrated to nearby farming villages and farming villages within Savannakhet Province, with few leaving the village for urban areas as was the case with the second generation (Table 4). The paddy area in Savannakhet Province is expanding, as noted below, and this moving out from the village played a major role in easing population pressures on the village.

If all the married couples who left the village had stayed, the paddy area per married couple would have been about 1.0 ha and only met 88% of the rice demand for each married couple family. Unlike the second generation, it was vital that the third generation move out from the village. However, even in the third generation, as noted above, there was surplus above the demand for rice for the married couples who remained in the village due to the many villagers who left.

4.2.4 Fourth Generation (Primarily Married Since the 1990s)

Because many in this generation were born since the 1980s, when the rapid population growth continued, there were 248 married couples more than those in the previous generation (Table 3), and 136 unmarried persons. Compared with the previous generation, the biggest change in this generation has been the sharp increase in circular migrants seeking better wages to Thailand since the 1990s, and the shift to a market economy following the "new economic mechanism" of 1986 (Rigg 2005: 20–24). In the 2000s, almost all young people in the village, regardless of gender, stayed in Thailand, particularly Bangkok, for periods of less than one year to multiple years, and some of them hardly ever returned to the village. They engaged in comparatively low wage occupations in Thailand, such as workers of small manufacturing, sales, and domestic help. The daily wage for agriculture in the village in 2014 was 30,000 Kip (US $3.7),[13] and while the daily wages in Bangkok varied according to occupation, the minimum was 300 Baht (US $9.1); thus, there was a wage difference of more than 2.5 times between the regions. As a result, young unmarried persons who left for circular migration were able to generate large amounts of money in a short time compared with the level available in Laos. They were able to buy motorbikes and mobile phones, whereas married couples built new modern houses, purchased durable goods, such as televisions and refrigerators, and obtained paddy fields. Since

[13] Foreign exchange rate according to the International Monetary Fund (2023).

the transition to a market economy, this expansion of consumption has spread, not only to the rural areas surrounding the villages but throughout Laos, although to varying degrees.

There were four major changes in livelihoods due to the circular migration to Thailand. Firstly, a class of people emerged in the village with a principal occupation of circular migration. They were primarily married couples in their 20s–30s, working in places such as Bangkok, who usually returned to their parent's home temporarily for the Laos New Year in April or for the busy farming season of rice planting or harvest and otherwise resided in Thailand. They built houses in the village to live in when they reached their 40s and finished circular migration, with some purchasing paddy fields.

Secondly, with an increase in circular migration, more than 30% of the entire population of the village in 2014 comprised temporary circular migrants to Thailand, as noted above. Their circular migration led to a reduction in village rice consumption. Most circular migrants stayed in Bangkok, so their meals were most likely from eating out, provided as board, or cooked by themselves with rice brought from the village. Although we do not know the accurate proportions for these forms of meals, according to some interviews, even if rice was brought, it was only when participating in circular migration. Assuming that circular migrants did not consume the rice produced in the village, it is possible to reduce the paddy fields by almost 35 ha. The progress in circular migration eased the population pressures.

Thirdly, as the population of the fourth generation increased, there was an increase in paddy field purchases due to circular migration. According to the results of the survey, the third generation purchased 42 ha of paddy fields between 2000 and 2014 from relatives, and people of the village and other villages. Cumulative purchased area in 2014 was more than double the purchased area of 39 ha up until 2000. The funds for such purchases were mainly from remittances derived from the circular migration of the fourth generation, and there was a high proportion of paddy fields acquired by their parents in the third generation. As of 2014, the age of females in this generation ranged from 15 to 40 years, which is the age group that bears the tasks of production in the village. However, as shown in Table 2, the proportion who had either inherited or purchased paddy fields was not yet very high. In 2014, excluding those undertaking circular migration, this generation comprised 50 married couple families residing in the village who owned 87 ha of paddy fields. i.e., each married couple family owned an average of 1.7 ha of paddy fields. The number of surviving children was 1.7, so the family of 3.7 persons produced at least paddy rice of 2.5 t, that is, 670 kg per person per year. This was adequate to live for the family. Naturally, these figures will change in the future, but assuming an increase in the purchase and inheritance of paddy fields, and the number of circular migrants remaining stable, any problems from population pressures are not expected to arise.

Fourthly, amid the growing population in the village and with the impact of persons engaged in circular migration who stayed, the proportion of permanent migration away from the village declined compared with the past (Table 3). As noted above, permanent migration to nearby farming villages and other rural areas has an important association with the population pressures on the village. Even in the fourth generation, the number of married couples increased with the population growth, and there was also a population of 136 unmarried persons. Once these unmarried people marry, the number of married couples in this generation will likely exceed 360.[14] This would result in more population pressures than those expected by the third generation. However, if we apply the current proportion of fourth generation circular migrants and permanent out-migrants leaving the village to unmarried persons, the large increase in married couples principally engaged in circular migration as well as an increase in the number of persons leaving the village means the number of married couples remaining in the village could approach 200. Even so, the number of people leaving the village remains at the same level and the role of out-migration remains important for this generation.

Why was it possible for the number of people leaving the village to migrate to nearby farming villages or farming villages in the province to grow? To stop the outflow of people from the village, there needs to be either an increase in paddy fields through the cultivation of new forest lands, as was the case when the village was first founded, or an increase in land productivity through multiple measures, including increased irrigation of existing paddy fields, the introduction of high-yielding varieties, and the use of fertilizers, pesticides, etc. In Savannakhet Province, the former reclamation of paddy fields had been dominant. In fact, the area of rain-fed paddy fields in Savannakhet Province as a whole increased from about 60,000 ha to 90,000 ha from the 1970s to late 1990s and then increased further to 170,000 ha in the 2010s (Lao PDR National Statistics Center 2005; Lao PDR Statistics Bureau 2015b). In Savannakhet Province, Central Laos, there remained forest land that can be used for reclamation and has enabled the expansion of rain-fed paddy fields for livelihood to absorb the population growth.

With regard to future increase of married couples, the number of births in the fourth generation has declined with the progress in family planning. According to interviews, most married couples are hoping to have 2–3 children. Furthermore, 95 of the 128 married couples in the fourth generation residing in the village report that they are using or have used contraception. As a result, there is unlikely to be as striking an increase in the number of married couples in the fifth generation as there has been to date. Although it is difficult to project the number of future married couples, we can make a simple estimate. As already mentioned, we assume the total number of married couples in the fourth generation that reside in the village may be 200. The average number of births for the fourth generation who married between 1995 and 2000 and are near completing fertility is 2.1. If we assume 75%[15] of the children marry someone from outside the village, the total number of married couples in the fifth generation will be about 350. This is about the same as the number of married couples from the fourth generation.

---

[14] Percentage of marriages in which the marriage partner is from outside the village is high, accounting for 76.5% of all marriages to date. Assuming that the entire unmarried population of 136 marries and 75% of these marriages are with people outside the village, these 136 future married couples together with the 248 already married couples result in a total of 367 couples for the number of married couples in the fourth generation.

[15] The assumptions for the calculations are the same as in Note 14.

How many of these married couples in the fifth generation will become village residents or move out from the village? This is also a difficult question. If we approach an era of families with two children, it will be necessary for a certain proportion to leave the village, such as through inheritance of land in other regions. In addition, if we consider other changes, such as whether circular migration to Thailand will continue to increase, purchases of nearby paddy fields will reach their limit, the employment markets in Laos will expand, and there will be progress in irrigation or an expansion of commodity crops as alternates to rice, the fifth generation is likely to follow a different path than the fourth generation.

5 Conclusion

This chapter clarified the response of livelihoods to the striking population growth through a fact-finding investigation using the case study of KH village; this village is a rain-fed paddy farming village in Central Laos that was founded in the 1920s–1930s and has had a rapid population growth since before World War II.

Here, we present the five outcomes, while also based on the five responses to the population growth described in Sect. 2. Firstly, despite the rapid population growth in the village that continued for a long time, the relationship between the population and the food generated from the livelihoods of the first to the fourth generation did not reach levels that would make it difficult for villagers to survive. There was no Malthusian response in which population growth exceeds survival resources, resulting in increased mortality due to poverty, etc. (Malthus 1798).

Secondly, regarding the population growth and land productivity, although further consideration is necessary, there was no irrigation even though chemical fertilizers were used. The rain-fed paddies in this village are subject to frequent floods and droughts, so the farmers must be aware of the benefits of increasing harvests and the costs to increase productivity from a long-term perspective. Thus, it is difficult to consider response from increase of land productivity as proposed by Boserup (1965).

Thirdly, permanent migration to other areas, particularly rural to rural migration, was a major response to the shortage of paddy fields due to the population growth. Most destinations were nearby farming villages and within Savannakhet Province. The area of paddy fields in Savannakhet Province increased. However, migration to other villages occurred even when there were surplus paddy fields for food. There was rural to rural migration not only for a response to the population growth, but also for improvements in so-called living standards.

Fourthly, circular migration to Thailand as temporary migration since the 1990s was also an important response to the population growth. Much of the income from circular migration by the fourth generation was used to purchase paddy fields. Finding new ways to generate cash income that coincided with the spread of the monetary economy to the village created a survival strategy unlike previous generations. Married couples with circular migration as their principal occupation and with virtually no dependence on paddy fields accounted for more than 60% of the married couples in this generation residing in the village. As a result, even though many had moved out from the village, the proportion of married couples remaining in the village increased. Among the people moving out from the village are circular migrants who marry those in other villages. These migrants were unlike those up until the third generation who moved out due to a demand for land.

Fifthly, the fourth generation was fully engaged in birth control through family planning. Many in the fifth generation who have been directly affected by this have not yet formed families, but they will undoubtedly experience a change from the increase in population up until the fourth generation. In addition, the response to livelihoods will likely be different from the fourth generation.

In summary, the response focused on the livelihoods due to the rapid population growth in KH village was unlike the Malthusian and Boserupian responses. The response was permanent migration among farming villages that have lower population densities and circular migration to Thailand as temporary migration.

Future issues are associated with declining birth rates due to conscious birth control since the late 1990s. Even in this village, there has been a pronounced decline in the population growth, particularly the natural growth rate, which is significantly impacting the population growth and livelihoods. Birth control is a problem associated with demographic transitions, but what types of factors are causing the recent decline in birth rate in Laos, including in this village, is an important research topic that is yet to be clarified. Unlike the demographic transitions in advanced nations, it will be interesting to examine the demographic transitions in Laos.

Acknowledgements This study was supported from JSPS Grants-in-Aid JP25257004. We sincerely appreciate the cooperation of the people in KH village with the investigation.

References

Bilsborrow R (1987) Population pressures and agricultural development in developing countries: a conceptual framework and recent evidence. World Dev 15:183–203. https://doi.org/10.1016/0305-750X(87)90077-5

Bongaarts J (1996) Population pressure and the food supply system in the developing world. Popul Dev Rev 22:483–503. https://doi.org/10.1007/978-3-031-11840-1

Boserup E (1965) The conditions of agricultural growth: the economy of agrarian change under population pressure. George Allen & Unwin, London

Carr D (2009) Population and deforestation: why rural migration matters. Prog Hum Geogr 33:353–378. https://doi.org/10.1177/0309132508096031

Coale AJ, Demeny P (1983) Regional model life tables and stable populations, 2nd edn. Academic Press, New York

Davis K (1963) The theory of change and response in modern demographic history. Popul Index 29:346–366. https://doi.org/10.2307/2732014

de Sherbinin A, Vanwey L, McSweeney K, Aggarwal R, Barbieri A, Henry S, Hunter LM, Twine W, Walker R (2008) Rural household demographics, livelihoods and the environment. Glob Environ Chang 18:38–53. https://doi.org/10.1016/j.gloenvcha.2007.05.005

Fukui H (1993) Food production in a Northeast Thai village. University of Hawaii Press, Honolulu

Halpern J (1961a) Population statistics and associated data in Laos. Laos Project Paper No. 3 (unpublished)
Halpern J (1961b) Laotian agricultural statistics. Laos Project Paper No. 9 (unpublished)
International Monetary Fund (2023) IMF data. https://data.imf.org/regular.aspx?key=61545850. Accessed 15 Feb 2023
Jayne TS, Chamberlin J, Headey DD (2014) Land pressures, the evolution of farming systems, and development strategies in Africa: a synthesis. Food Policy 48:1–17. https://doi.org/10.1016/j.foodpol.2014.05.014
Kuchiba M (ed) (1990) Don Daeng mura no dentōkōzō to sono henyō (Traditional structure of don daeng village and its transformation). Sōbunsha, Tokyo (In Japanese)
Lao PDR, Statistics Bureau (2015a) The results of population and housing census. https://laosis.lsb.gov.la/board/BoardList.do?bbs_bbsid=B404. Accessed 23 Oct 2022
Lao PDR, Statistics Bureau (2015b) Statistical yearbook 2015, Vientiane
Lao PDR, Statistics Bureau (2015c) Statistics 1975–2015, Vientiane
Lao PDR, National Statistics Center (2005) Statistics 1975–2005, Vientiane
Malthus TR (1798) An essay on the principle of population: an essay on the principle of population, as it affects the future improvement of society with remarks on the speculations of Mr. Godwin, M. Condorcet, and other writers. Printed for J. Johnson, St. Paul's Church-Yard, London
Mizuno K (1981) Tai nōson no shakai soshiki (Social structure of Thai farming villages). Sōbunsha, Tokyo (in Japanese)
Mwesigye F, Matsumoto T, Otsuka K (2017) Population pressure, rural-to-rural migration and evolution of land tenure institutions: the case of Uganda. Land Use Policy 65:1–14. https://doi.org/10.1016/j.landusepol.2017.03.020
Nakamura S, Iwasa M (2008) Permeation and history of modern medicine in Laos. In: Daniels C (ed) Collected papers on eco-history of monsoon Asia, vol 2. Kōbundo, Tokyo, pp 229–246 (in Japanese)
National Institute of Population and Social Security Research (2022) Collection of population statistics. https://www.ipss.go.jp/syoushika/tohkei/Popular/P_Detail2022.asp?fname=T05-02.htm. Accessed 8 Jan 2023
Niwa T, Nakagawa S, Takahashi S, Nishimoto F (2021) Transnational labor migration from rural village in Laos to Bangkok, Thailand: focusing on relationship with the home village. Nenpō Tai Kenkyū (J Thai Stud) 21:17–30 (in Japanese)
Rigg J (2005) Living with transition in Laos: market integration in Southeast Asia. Routledge, London
Statistics Bureau, Prime Minister's Office, Japan (1963) Nihon no jinkō: shōwa 35 nen kokusei chōsa no kaisetsu (Japan's population 1960: explanation of the 1960 national census), Tokyo (in Japanese)
Stuart-Fox M (1997) A history of Laos. Cambridge University Press, Cambridge
Takahashi S, Yokoyama S, Nishimoto F, Niwa T (2023) Population growth and generational response to livelihood choices in a rain-fed farming village in central Laos. E-J GEO 18:264–278 (in Japanese). https://doi.org/10.4157/ejgeo.18.264
United Nations (2022) 2022 Revision of world population prospects. https://population.un.org/wpp/Download/Standard/MostUsed/. Accessed 10 Nov 2022
World Bank (2023) GDP growth (annual %), Least developed countries: UN classification. https://data.worldbank.org/indicator/NY.GDP.MKTP.KD.ZG?end=2021&locations=US%29-XL&start=1980&view=chart. Accessed 22 April 2023

Transnational Labor Migration from Rural Village in Laos to Bangkok, Thailand

Focusing on Relationship with the Home Village

Takahito Niwa, Shinichi Takahashi, and Futoshi Nishimoto

Abstract This chapter investigates whether network ties at the village level support transnational labor migration from Laos to Thailand. The research focused on individual-level and household-level movements of migrants in interview surveys in KH village, central Laos located near the border, and in Bangkok, Thailand. The surveys in Bangkok were conducted in September 2014, March 2015, and November 2016. We found that the lack of employment opportunities in Laos and the low barriers to emigration to Bangkok was a common cause of transnational labor migration. The development of ICT helped villagers to obtain an image of life and work in Bangkok before they depart. The majority of households in the village had at least one member working in Bangkok. As a result, the in situ population of KH village has begun to decline. In Bangkok, Thai employers obtained the stay and work permit for the migrants as part of their compliance. Even so, working conditions have not sufficiently improved, and migrant workers are using local networks in KH village to avoid the risk of poor treatment. The people of the village also maintained a support network for migrant workers. We also found close financial ties between the village and Bangkok. Most remittances were spent for daily living expenses rather than to invest in the future.

Keywords Transnational migration · Network tie · Remittance · Laos · Thailand

---

This chapter was a revised version of Niwa et al. (2021).

T. Niwa (✉)
College of Human and Social Sciences, Kanazawa University, Kanazawa, Japan
e-mail: niwa-2@staff.kanazawa-u.ac.jp

S. Takahashi
Kobe University, Kobe, Japan
e-mail: takahashi@econ.kobe-u.ac.jp

F. Nishimoto
Embassy of Japan in Laos, Vientiane, Laos
e-mail: nsmtfts@gmail.com

S. Yokoyama et al. (eds.), *Population Dynamics and Livelihood Changes of Small-Scale Societies in Laos*, International Perspectives in Geography 22,
https://doi.org/10.1007/978-981-96-2956-5_5

1 Introduction

1.1 Research Background

Many researchers have been arguing that population growth in rural areas drives rural–urban migrations (Pham and Hill 2008; Qin 2010; Jayne et al. 2014). In Laos, where agriculture occupies 26% of GDP, the rural population growth is also driving a shift from traditional subsistence agriculture to a variety of livelihood strategies, including commodity crop cultivation and migrant work (Rigg 2007; Rigg et al. 2016; Cole and Rigg 2019).

Laos transitioned to a socialist system in 1975, which brought an end to the domestic turmoil associated with the civil war. Parallel to this, there were social developments such as a spread of public health throughout the country, a decline in the birth rate, and an increase in average life expectancy (Jones 2013; Gubhaju 2007). At the same time, however, the manufacturing sector is still underdeveloped. The Economic gap with Thailand reached 2.7 times in GDP per capita as of 2015.[1] Due to limited domestic employment opportunities, there is an asymmetry in movement of goods from Thailand to Laos and people from Laos to Thailand (Chanthavysouk 2006). Migration from areas near the border to Thailand is also remarkable (Huguet and Punpuing 2005; Sciortino and Punpuing 2009; Rungmanee 2016). Compared to Burmese and Cambodians, Laotians are similar to Thais in terms of language and culture. They work as domestic workers rather than migrants from two countries (Pholphirul and Rukumnuaykit 2010). Nevertheless, because Lao workers are foreigners in Thailand, they have to deal with inhumanity and institutional gaps.

In this chapter, therefore, we show that village networks reduce barriers and risks to migration to Thailand using a case study of a rural village in central Laos.

1.2 Previous Studies

Phouxay (2017) pointed out that migrant workers from Laos to Thailand decided to move at the individual level to escape poverty. Manivong et al. (2014) noted that cash earning through migration has become one of the most important livelihoods in rural areas in Laos. There is a correlation between household size and the number of migrant workers. While this indicates that the households decide to migration work as part of a livelihood strategy, rather than at the individual level, the impact of networks to migration at the village level on the decision has not been examined. Brees (2010) has shown that Burmese shape ethnic networks in Thailand, which function to support their employment and daily life. However, these studies do not indicate whether ethnic networks are related to the migrants' home villages since they are based on research in Thailand.

[1] GDP per capita in Laos as of 2015 was 2,125 U.S. dollar.

On the other hand, while transnationalism and transnational communities are becoming the new analytical framework for international migration studies, this mainly focuses on the construction of de-nationalized migration networks (Castles et al. 2014) and does not approach local communities. In other words, the most important networks for migrants may be very local ones, such as family and local ties. It has not been clarified whether local ties support cross-border migration. It can be hypothesized that migration may minimize risks not only at the individual or household level but also through networks at the village level.

1.3 Research Objectives and Methods

In this chapter, we clarify whether family and local ties among people at the village level support migration through an examination of information on human relations and information among migrants focusing on migration from rural Laos to Thailand. To achieve this purpose, we investigated the role of local networks between the home village and settlements in Thailand in relation to their work, daily life, and remittances to the village, after obtaining an overall picture of migration in rural Laos.

Our research methods are as follows. We conducted a structured interview to all households in KH village in central Laos, and to 135 villagers who are migrant workers in and around Bangkok, Thailand.[2] The survey was conducted via a questionnaire prepared in English with a Lao-English interpreter. In order to identify population changes in the village, we conducted unstructured interviews in Lao to changes in household composition over time, as well as to track and record absentees who once lived in the village and have since died or moved out (Nishimoto 2011). We were able to trace back to people who were born in the late nineteenth century. However, it was difficult to trace the further back, and this made the more inaccurate the information. Therefore, we decided to use only data from 1970 onward, when the memories of people in their 60s and 70s who still live in the village are considered to be reliable.

The networks used in this chapter are defined as informal social networks formed by migrants themselves to facilitate their mobility (Castles et al. 2014). Such social networks have been discussed as migrant network theory and migrant systems theory (e.g., Massey et al. 1993; de Haas 2010), and chain migration has also been used in this context. Furthermore, we discuss as dividing social networks into family networks and local networks. The family networks are the social network among relatives within a KH village based on information on changes in household composition over

[2] Prior to the survey to migrant workers in Bangkok, we obtained contact information from relatives of migrants in KH village and contacted them in Bangkok. In addition, we had some villagers such as village head accompany us to contact between KH villagers in order not to make migrant workers feel insecure. The surveys in Bangkok were conducted in September 2014, March 2015, and November 2016.

time. The local networks are the social network among acquaintances and friends in KH village who are not relatives.

In the following section, we will summarize the overview of the migrant works in KH village in Sect. 2. Next, we analyze the characteristics of the migrant works heading to Thailand in Sect. 3. Finally, in Sect. 4, we examine the relationship between the migrants and the village.

1.4 Circumstances for Low-Skilled Foreign Workers in Thailand

At first, we would like to briefly summarize the circumstances surrounding low-skilled workers in Thailand. Rural surplus labor was absorbed by population migration between rural areas until the 1970s in Thailand. Since the late 1970s, economic growth has led to massive rural–urban migration which has been absorbed by employment in the informal sector and the industrial sector (Goldstein and Goldstein 1986).

The Thai government has restricted the intake of low-skilled workers from neighboring countries since the 1930s (Chalamwong 2011). However, as the birth rate declined and the economy grew rapidly in Thailand, Thai workers became less willing to take low-wage labor and 3D jobs (Chalamwong and Prugsamatz 2009: 5), and labor shortages have begun to emerge. Since the late 1980s, unskilled workers from neighboring countries have been smuggled into the country. Thai government quietly accepted the existence of illegal foreign workers while upholding the principle of not allowing foreigners to enter and work illegally (Kaur 2010). The number of foreign workers from the CLM (Cambodia, Laos, and Myanmar) countries neighboring Thailand is estimated to be about 1.8 million as of 2007 (Martin 2007), many of whom are irregular migrants without passports or other documents (Huguet and Punpuing 2005). The working and living conditions in which irregular migrants are placed are harsh (Molland 2010), and they are also victims of police rip-offs and exploitation by employers due to their lack of identification.

Since the 1990s, the Thai government has been coping with the CLM countries on the bilateral Memorandum of Understanding on Cooperation in the Employment of Workers and the Alien Working Act (2008)[3] has begun to accept low-skilled workers by gradually opening the door for legal stay and legal work to foreigners from neighboring countries who are in the country irregularly (Skeldon 2006: 285).

[3] While the Royal Ordinance Concerning the Management of Employment of Foreign Workers came into effect in 2017 and there are changes in the management system for foreign workers, this chapter is based on earlier survey data and does not address this point.

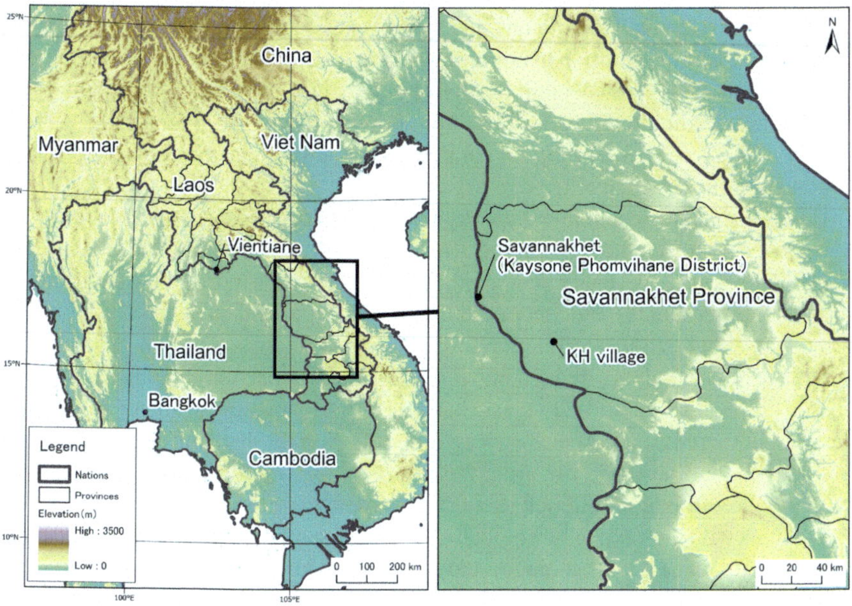

Fig. 1 Study area

2 Migrant Workers in KH Village

2.1 *KH Village Overview*

KH village as a case study is located in the western part of Savannakhet Province in central Laos, approximately 70 km from Savannakhet City, the provincial capital (Fig. 1). The Mekong River flows southward at the western edge of the province, which forms the border with Thailand.

KH village is an administrative village consisting of one settlement of the ethnic Lao[4] with 112 households[5] and 760 people as of the end of 2014.[6] The main livelihood is rice cultivation in rain-fed paddy fields. Paddies are often flooded during the rainy season making the harvest unstable because some of the rain-fed paddy fields are in low land. Cattle and chickens are raised by many households and are regarded as assets to be cashed in the future, in addition to for private consumption. Some

[4] The ethnic Lao generally live with the wife's parents after marriage. The youngest daughter often takes care of her parents, and females often inherit property from their parents (Thomson and Baden 1993).

[5] A common household is composed of a married couple or parents and children (Dommaraju and Tan 2014).

[6] Surveys in the village were conducted in September 2013, December 2013, and September 2014. It should be noted that the timing differs from the survey in Bangkok and there is some variation in the meaning of the figures. Note that only age was fixed as of the end of 2014.

households also grow watermelons and other vegetables as cash crops to generate cash income. This, in addition to migrating to Thailand, creates a complex livelihood. In terms of household ownership of household appliances in the village, television was the most common (87%), followed by mobile phone (85%), motorbike (78%), and tiller (74%) in 2013. The number of households with televisions and refrigerators is increasing in KH village.

Here we show the economic disparity in the village as a result of the survey of 99 households[7] conducted at the end of 2013. The majority of households spend less than 300,000 kip (1,212 baht)[8] per month, while some spend more than 1,000,000 kip (4,040 baht) per month. The amount of household expenditures depends on the size of the household, but the correlation is small. Based on the amount of expenditure, the village households are divided into quartiles, and comparisons among the quartile classes (Table 1).[9] There is no clear difference in the area of farmland among the classes, but there is a difference in rice yield. It is thought that land conditions, land productivity, and pesticide use are affected combined. As for livestock ownership, the only class in which differences among classes were observed was the buffalo. Households with high expenditures tend to own a larger number of buffalo which have high property values. On the other hand, all households owned more than one cow, regardless of their economic circumstance. The percentage of temporary migrants[10] among the household members shows little difference among classes. This means migrant works are a common feature of the village.

We introduce two examples of households that may illustrate the differences between village classes. The first is headed by a husband born in 1968 and a wife born in 1967 and has a household of 13 members. This household is in the highest expenditure class (Q4) in the previous household hierarchy. They have seven children, one of whom died prematurely. Three children have temporarily left the village, two to Bangkok and one to the City of Savannakhet. They engage in wet-rice agriculture on their paddy, 2.5 ha, inherited from the wife's parents, and harvest about 3 tons a year. They don't sell rice. They also grow watermelons and cucumbers as cash crops and earn about 20,000,000 kip (80,800 baht) a year. Furthermore, they receive the same amount in remittances from Bangkok.

[7] The number of households differs from the previous 112 because there are 13 households in which only the elderly or children reside. These households have no dependents living in the household due to migrant workers or bereavement. They share meals and farm work with their relatives and were interviewed together during the all-inclusive survey.

[8] The exchange rate between the Laotian kip and the Thai baht, was converted at 1,000 keep = 4.04 baht (2014 average). The exchange rate between the U.S. dollar and the baht, discussed below, was also converted at 1 dollar = 32.5 baht for the same period (https://fxtop.com/, last accessed February 12, 2021).

[9] We used the amount of expenditures instead of household income although the income is more suitable for analysis of household class. The reason that we asked the expenditure was that villagers feel easier to answer in the survey and that it is possible to be connected to statistical data such as household surveys. In addition, the number of pigs and poultries were excluded from the analysis.

[10] Temporary migration from the village was defined as any movement other than leaving the village permanently by changing one's family register, with the intention of returning to the village. Migrant works are also included in this category.

Table 1 Comparison of population, farmland, and livestock per household group classified by expenditure classes in KH village (n = 99)

	Quartile classes divided by monthly expenditure (thousand kip)				Average	ANNOVA/ K-W test *p*-value	Difference among classes
	Q1 (<=300)	Q2 (301–400)	Q3 (401–500)	Q4 (>=501)			
	Very poor	Poor	Rich	Very rich			
	n = 44	n = 11	n = 22	n = 23			
Population							
The number of household members	6.18	8.36	7.36	8.35	7.18	*	Q1 < Q4
The number of temporarily migrated members	1.84	1.73	1.73	1.61	1.75	n.s.	
Rate of temporary migrants (%)	28.2	20.4	20.2	19.5	23.6	n,s,	
Farmland							
Irrigated paddy fields (ha)	0.06	0.27	0.05	0.80	0.25	n.s.	
Rain-fed paddy fields (ha)	1.52	1.30	1.85	2.08	1.71	n.s.	
Swidden (ha)	0.12	0.62	0.20	0.24	0.22	n.s.	
Rice yield (ton)	1.91	1.75	2.68	2.79	2.27	**	Q1 < Q4
Livestock							
The number of buffalos	0.43	0.36	1.23	1.30	0.80	*	Q1 < Q3

(continued)

Table 1 (continued)

	Quartile classes divided by monthly expenditure (thousand kip)				Average	ANNOVA/ K-W test *p*-value	Difference among classes
	Q1 (<=300)	Q2 (301–400)	Q3 (401–500)	Q4 (>=501)			
	Very poor	Poor	Rich	Very rich			
	n = 44	n = 11	n = 22	n = 23			
The number of cows	3.98	5.91	6.18	6.00	5.14	n.s.	
The number of goats	0.00	2.27	0.00	0.17	0.29	n,s,	

Note For each item, the Bartlett test was used to confirm whether equal variances were found, followed by ANNOVA if equal variances were found, and the Kruskal–Wallis test if not. For items for which significant differences were found, TukeyHSD test was conducted after the fact for multiple comparisons when equal variances were found, and Steel–Dwass test was conducted when not found. ** indicates a significant difference $p < 0.05$ and * indicates a significant difference $p < 0.1$

Source Authors' interview survey

The other household is in the lowest expenditure class (Q1) in the household hierarchy and is headed by a husband born in 1963 and a wife born in 1966. They had three children, two of whom died about a month after birth. The one remaining child got married in 2014, and he left his wife and children to go to Bangkok to work. This household inherited a little less than 2 ha of farmland from the husband's parents and gets about 4 tons of rice a year. They sell a part of harvests. They also earn about 4,000,000 kip (16,160 baht) a year from growing watermelons and raising poultry. All of the household's income comes from farming and they do not receive remittances from their children.

2.2 *Expansion of Migrant Work*

The population in KH village, as reconstructed from the interviews (Fig. 2), had been continuously increasing until around 2000. Recently, the population is in a state of stagnation due to the declining birth rate. The resident population, which is the population minus those who temporarily left the village, peaked at 640 in 2000 and has been declining, reaching 561 as of 2014.

On the other hand, the number of temporary migrants continues to increase. The earliest recorded temporary migration from KH village within the Lao PDR was in

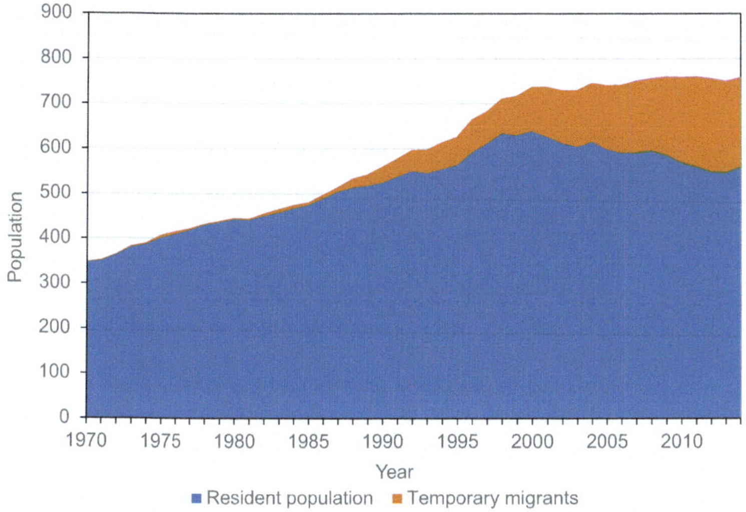

Fig. 2 Population change in KH village, 1970–2014. *Source* Authors' interview survey

1971. Temporary migration to Bangkok[11] began to increase in the 1990s. In 1990, 35 people from KH village temporarily migrated to Bangkok, which was only 6% of the population at the time. By 2014, 199 people, or 26% of the population, had left the village to move to Bangkok. Eighty-four households, or 75% of the 112 households in the village, have household members who are in Bangkok. In these households, an average of 2.4 people, or 36% of household members, have left the village.

The average age of the 11 new temporary migrants in 1990 was 15.2 years old, the average age of the 13 migrants in 2000 was 18.6 years old, and the average age of the 22 migrants in 2010 was 16.7 years old. The percentage of those who were under 20 years old when they moved out from the village has been consistently high, hovering around 80%.

In some cases, the decision to migrate is part of a household livelihood strategy in which parents and children discuss to determine the migration from within the household. In other cases, the decision is made by the child themselves to leave the village without informing his or her parents. Bangkok is no longer a distant unknown destination for teenagers in particular, as even in the KH villages they have access to information about the city via satellite TV, mobile phones, and especially social networking services.[12]

[11] KH villagers regarded the distant areas of Thailand as Bangkok that were not around the border. The survey in Bangkok revealed that migrant workers were actually scattered over a wider area than the Bangkok metropolitan area. We refer Bangkok as same as villagers recognize.

[12] Some households note the names of social networking accounts of their children who are migrating to Bangkok to work.

Table 2 Working industries of migrant workers in Bangkok by sex and age group (n = 135)

Sex/age groups	Primary industry	Secondary industry	Tertiary industry	Total
Male	11	25	46	82
10–19	2	6	22	30
20–29	5	10	23	38
30–39	3	7	0	10
40–	1	2	1	4
Female	7	19	27	53
10–19	1	3	8	12
20–29	1	10	12	23
30–39	4	4	3	11
40–	1	2	4	7
Total	18	44	73	135

Source Authors' interview survey

3 Departure Migration to Bangkok

3.1 Overall Picture of Migrant Workers

First, we review the basic demographics of the survey informants in Bangkok. We interviewed 135 migrants from KH villages between 2014–2016.[13] The sex of them was slightly more male than female, with 82 males and 53 females (Table 2). The majority were young, with 42 in their teens, 61 in their 20s, 21 in their 30s, and 11 in their 40s or older.

Of the 135, 58 are unmarried, 73 are married couples, and 4 are divorced. Of the 73 married couples, 30 are couples who are both from KH villages. Of the total 58 married couples, 37 have children and the number of children is 61. The residence of the children varies by household. There are 11 households with parents and children living together in Bangkok, and the number of children is 25, accounting for 41% of the total. Most children living in Bangkok work with their parents or stay at home even though they are in age in school. Some children attend elementary school through the intermediary of their Thai employers. On the other hand, the number of children living with grandparents or relatives in KH Village is 21 (34%), as skipped

[13] The survey in Bangkok is conducted by a snowball sampling based on information provided by villagers, and the possibility that it is biased toward information from specific households cannot be ruled out. However, the data are sufficiently debatable because they can be linked to information on the KH village side. Of the 135 informants, 44 were not counted as temporary migrants as of 2014. 26 were resident in the village at the end of 2014 and subsequently moved to Bangkok. 17 were permanent migrants, and one was from another village who married a temporary migrant from KH village in Bangkok. Those 18 were related to KH village through sending remittances to their relatives in KH village at the time of the Bangkok survey.

generation households.[14] The situation is similar to that of Thai migrant workers, who rely on support from relatives and continue to work during their child-rearing years (Podhisita 2017).

The industries in which migrants from KH villages work are diverse. There are 18 in the primary industry, 44 in the secondary industry, and 73 in the tertiary industry. Typical places of employment include shrimp farms in the primary industry, garment factories in the secondary industry, and restaurants in the tertiary industry. Most of them are engaged in low-skilled and labor-intensive jobs without job titles. This is due in part to the low level of education among migrant workers. The average years of schooling for migrant workers is 4.5 years, which is about the same as a primary school degree.[15] The younger the age, the longer the years of schooling, but even among those in their teens and twenties, the average year of schooling is only 4.9 years.

3.2 *Chain Migration Through Family and Local Networks*

Villagers, including young teenagers, are able to come to Bangkok to work because of the existence of a chain migration network mediated by family and local ties within KH village. Both the family network mediated by relatives, and the local network mediated by acquaintances and friends in the village, support the migrant workers in Bangkok.

The case of Mrs. Khan (pseudonym) is representative of the family network. Her family can be traced back to the founding of KH village, and her father was the village head at the time of the study. 42 years old Mrs. Khan was the first person who came to the shrimp farm located near the Gulf of Thailand from KH village after receiving information from her husband's relatives about the farm. Then, she passed on the information to her brother-in-law, and villagers moved here one by one. All 18 people who work in the primary industry are employed at this shrimp farm. All of them live in free accommodation on the farm.[16] Since the migrants do not have access to transportation outside the farm, their Thai employers take them to a nearby retail store once a month to help them buy daily necessities. The migrant workers did not complain of any inconvenience in this circumstance but were rather satisfied with their work and life here. Informants say that it is similar to the atmosphere of the KH village without the hustle and bustle of the city. They are also satisfied that they can reduce spending and save and transfer a lot of money. The average age of the 18 people working here is 29.8, much older than in other places of employment

[14] The remaining 15 children are those living in other villages in Laos (10) or separated from their parents in Thailand (5).

[15] The duration of primary education in Laos is five years and secondary education is six years. Primary education is six years and secondary education is also six years in Thailand.

[16] Their employers also pay for food and utilities, so migrant workers pay no living expenses.

in Bangkok. This also indicates that they are oriented toward long-term employment at the shrimp farm.

Mrs. Khan has already left the farm to work in a large clothing market in Bangkok. Sixteen migrants from KH village, including her children, work in the market as salespeople and construction workers. The introduction of jobs and housing by Mrs. Khan has strengthened the villagers' desire to migrate. As social ties of women in their home community are one of the factors that encourage migration (Thongyou and Ayuwat 2005), Mrs. Khan is at the center of social ties related to migration in the village. She first shares information with her relatives and then supports other villagers' migration. It is possible to say that her high social status in the KH village also contributes to the reliability of information about Bangkok. This suggests that social relations in her village are involved in the formation of local networks outside the country.

A local network mediated by acquaintances and friends from the same KH village can be shown at 24-hour restaurants in the city of Bangkok. Twenty-six villagers were identified as working in the same restaurant chain. In this restaurant chain, the upper floors of the restaurant are used as living space for the employees, who pay 500 baht per person per month. The average age of the 26 workers is 19.8 years old. Compared to the shrimp farm, migrants get information mainly from friends and acquaintances in KH village. Of the 25 people interviewed about sources of information about the restaurant, 9 decided to work here based on information from relatives, and 9 based on information from friends and acquaintances. The remaining seven were cases in which they had received information from friends but were unable to distinguish whether inside or outside the village. The local network is formed when villagers who have known each other since childhood exchange information with each other.

Mr. Phet (pseudonym), who has been working here for a long time and is the only villager who holds the position of supervisor, plays a significant role in this local network. Mr. Phet started working at this restaurant chain after receiving information from a friend in the village, then gave information to his friend Mr. Vieng (pseudonym) in the village. Mr. Vieng also moved here to work. Similarly, Mr. Sai (pseudonym) moved here after receiving information from a friend in the village and gave information to his younger brother. Since young people of the same generation work in the same store here, they can work and live while maintaining friendships within the KH village. This is what attracts the youth of KH village to this restaurant chain.

However, not all of the migrant workers in Bangkok take advantage of the family and local ties in the KH village. There are cases of independent stay and work without the social network, such as domestic workers who live and work in their employers' houses.

3.3 Improvement of Work and Life Circumstances in Bangkok

As of 2015, Lao nationals entering Thailand newly are required to have a passport. Since it is possible to stay and work legally in Thailand by obtaining a work permit document even if they have been staying and working illegally (Otomo 2018), the number of Lao workers who move to Thailand legally is increasing (Huijsmans and Phouxay 2008; Huijsmans 2014: 339–343).

Even among migrants from KH villages, 106 out of 135 have some kind of documentation, and many have already switched to legal employment.[17] These legalization procedures are driven by Thai employers. This is thought to be due to business decisions by Thai employers as a means of attracting workers to address labor shortages and as a means of compliance in response to regulations on illegal workers by the Thai government. In some cases, the employer bears the entire cost of the legalization procedures, while in other cases, the migrant workers from the KH villages bear the cost. The median cost they paid to obtain the certificate was 3,000 baht.

The average monthly income of migrants from KH villages is 8,484 baht (Fig. 3).[18] This is higher than the 7,800 baht per month for those working 26 days per month at the minimum wage of 300 baht per day (one day off per week) in Thailand as of 2015. The average amount for those without a work permit is 7,850 baht, while the average amount for those with a work permit is higher at 8,680 baht. This is consistent with the discussion that regular migrants earn more than irregular migrants (Holumyong and Punpuing 2014). However, nearly half of the informants are still forced to work for less than the minimum wage. While many of these workers earn in the 7,000 baht range, one earns only 3,000 baht per month. It is a fact that Lao migrant workers are in a vulnerable work circumstance.

3.4 Remittances to KH Village

Migrants from KH village also send substantial amounts of remittances to the village, as they move to Bangkok mainly earning cash. 105 people who send remittances to KH village send 4,048 baht per month, compared to an average income of 8,474 baht for them.[19] The proportion reaches 48%.

The local network also functions as a way to send money to KH village, where Mr. Thon (pseudonym) owns a private car and a Thai bank account, and migrants transfer money to Mr. Thon's account first. They then inform Mr. Thon by mobile phone of the amount of money transferred, and Mr. Thon goes to Mukdahan City to

[17] On the other hand, four migrant workers from KH village did not have a work permit or a passport.

[18] Some informants answered not their individual income but their household income. The per capita income was divided by the number of people working in the household in such cases.

[19] We checked of the amount received by the 20 households in KH Village. They receive remittances from Bangkok an average of 4,000 baht per month as well.

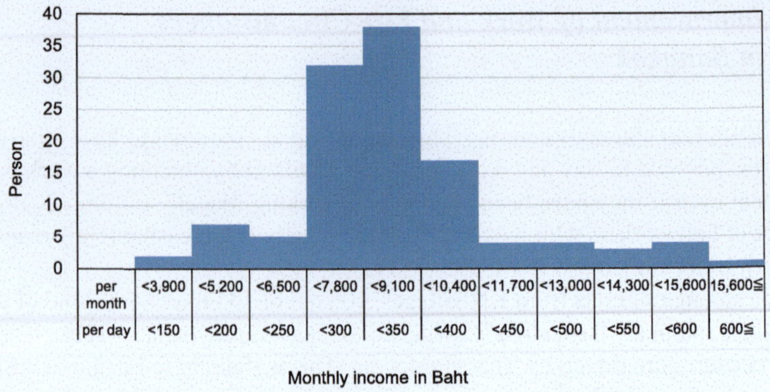

Fig. 3 Histogram of the monthly income of migrant workers in Bangkok, 2014–2016 (n = 117). *Note* The daily rate is the monthly amount divided by 26 days. *Source* Authors' interview survey

withdraw cash. Mr. Thon also runs a retail business in the village, so he also buys goods and provides an intermediary service for international money transfers. Mr. Thon receives a commission of 3% of the remittance amount.

The cash sent to the village supports its monetary economy. According to interviews with 20 households with temporary migrants in Bangkok, the cash was used to supplement the household's livelihood (12 households), to build a new house (10 households), and to take care of children (6 households) as multiple responses. Cash from migrant workers is used to invest in real estate and for consumption in daily life rather than invest in farmland in KH village.

4 Relationship Between Migrants and the Village

Migration from KH village to Bangkok is already a common practice seen in the majority of households. Because of the importance of earning cash through migration, many younger generations are migrating, and the number of skipped generation households is increasing. In particular, the large number of those who migrate in their teens suggests that they prioritize work opportunities over educational opportunities. In KH village, it is possible to say that the immediate gains from migrant work are considered more important than investment in children's education. One factor in this decision-making process is the limited employment opportunities in Laos even though workers with higher education. In addition, information about Bangkok prior to migration may have facilitated migration, as the psychological distance has been shortened by the diffusion of ICT networks, which has lowered the barriers to migration to Bangkok.

Many migrants from KH village choose to move to places where villagers are staying and working and where they can have a job and shelter at the same time. Since the problem of illegal employment and human trafficking are also vulnerabilities and risks for Lao people (Phouxay 2017; Huijsmans 2007, 2014), migrants from KH village have established a village-level local network in Bangkok as well and have developed reliable information, they are likely making decisions to minimize travel costs and risks to Bangkok. Our survey data suggest that the formation of these networks took shape in the 1990s.

The income of migrant workers in Bangkok may reflect this risk. Normally, the institutional framework of legal work procedures should play a role in reducing their risk, however, as Mori (2009) points out, the circumstances of work and life have not improved. While some migrant workers earn higher incomes with a passport or a work permit, many others earn below average wages. In order to avoid these risks as much as possible, many migrant workers are thought to use networks within KH village.

The large remittances from the migrant workers suggest that migration has become a significant part of KH village's livelihood. Compared to Southichack (2013), who estimated the reasonable amount of remittances for Lao workers working in Thailand to be 553–624 dollars per year (1,498–1690 baht per month), it can be inferred that the cash earnings from migrant workers in KH village have a very large impact to the village. The large remittances to KH village can be attributed to the thrifty lifestyle in Thailand. As in the case of the shrimp farm and the restaurant chain, the migrants kept their consumption expenditure to a minimum. The local network of KH village also functions for international remittances. It can be said that the local network in KH village comprehensively supports migration from the village to Bangkok, work and life in Bangkok, and remittances from Bangkok.

In Laos, farmland is divided equally among all children upon inheritance, so in times of population growth, it is not possible to secure enough farmland to live on (Tomita et al. 2015), and it is necessary to purchase use rights to farmland inside and outside the village or to invest in livelihoods (Manivong et al. 2014). In the case of KH village, a reluctance to acquire farmland was observed at the time of the interview, while on the other hand, house construction was active. This is both an investment in the future life of the temporarily migrated villagers and supports for economic cycle of the village through a part of the construction fee that goes to the villagers who participate in the construction works. Since the residential population of KH village was declining at the time of our surveys, villagers didn't consider the inheritance of farmland to be a problem. It is probable that some of the migrant workers save up to acquire the right to use farmland, and the problem of the shortage of farmland or the right to use farmland will become more apparent in the future.

5 Conclusion

This chapter focuses on transnational migration from rural Laos to Thailand and examines whether village-level villager connections support migration. The large number of migrant workers from KH village to Bangkok is due to both the lack of employment opportunities in Laos and the low barriers for villagers migrating to Bangkok. The development of ICT allows villagers to have a picture of life and work in Bangkok prior to migration and has lowered the psychological barriers to cross-border migration. The majority of households in the village have temporarily lost their household members for migrant works, thereby entering a phase of decline in the village's resident population.

On the other hand, Thai employers in Bangkok handle the legalizing procedures of Laotian employment. This also lowers barriers to migration from an institutional aspect. The family and local networks have been established among KH villagers to support migration to Bangkok, from the information on employment opportunities in the pre-migration stage to international remittances in the post-migration stage. In addition, the village and Bangkok are deeply connected financially, and it can be pointed out that there is more international economic circulation between Thailand and Laos than has been assumed previously (Southichack 2013). Remittances to Laos are consumed in the villages for current expenses rather than for investment in the future.

It can be broadly argued that, in the context of the scale of labor migration from Laos to Thailand, migrant workers play an important role in rural livelihoods in central Laos, close to the Thai border, even though this chapter is positioned as an exceptional case in terms of the amount of remittances sent. However, the fact that not all migrants move through family and local networks suggests that relationships and interests in rural areas have a more complex structure. Since we lack sufficient research data to discuss this point, we would like to consider it in the future.

Acknowledgements We would like to thank the people of KH village for their generous cooperation in conducting this study. We also thank Ms. Sychanh Pasansyms, a district health volunteer for coordinating with villagers during the interviews. This study was supported from JSPS Grants-in-Aid JP25257004. The field survey was conducted by the National Institute of Agroforestry of Laos and the National Institute of Public Health of Laos (now the Lao Tropical and Public Health Institute) as counterpart institutions. Finally, this chapter is incomplete without a specific acknowledgement of the contributions of late Prof. Satoshi NAKAGAWA passed in 2019. We dedicate this to him.

References

Brees I (2010) Refugee and transnationalism on the Thai-Burmese border. Global Netw 10:282–299. https://doi.org/10.1111/j.1471-0374.2010.00286.x

Castles S, de Haas H, Miller MJ (2014) The age of migration: international population movements in the modern world, 5th edn. The Guilford Press, New York

Chalamwong Y (2011) Different stream, different needs, and impact: managing international labor migration in ASEAN: Thailand (Emigration). PIDS Discussion Paper Series 2011–27

Chalamwong Y, Prugsamatz R (2009) Labor migration in Thailand: recent trends and implications for development. TDRI Quart Rev 24(3):3–9

Chanthavysouk K (2006) Labour export: a contribution to Lao development. Technical Background Paper for the third National Human Development Report Lao PDR 2006

Cole R, Rigg J (2019) Lao peasants on the move: pathways of agrarian change in Laos. Aust J Anthropol 30:160–180. https://doi.org/10.1111/taja.12312

de Haas H (2010) The internal dynamics of migration processes. J Ethn Migr Stud 36:1587–1617. https://doi.org/10.1080/1369183X.2010.489361

Dommaraju P, Tan J (2014) Households in contemporary Southeast Asia. J Comp Fam Stud 45:559–580. http://www.jstor.org/stable/24339641

Goldstein SE, Goldstein A (1986) Migration in Thailand: a twenty-five-year review. Papers of the East-West Population Institute, no 100

Gubhaju B (2007) Fertility decline in Asia: opportunities and challenges. Jpn J Popul 5:19–42

Holumyong C, Punpuing S (2014) A cost-benefit analysis of the legal status of migrant workers in Thailand. In: Adams RH, Ahmad A (eds) Managing international migration for development in East Asia. World Bank, Washington, DC, pp 263–282

Huguet JW, Punpuing S (2005) International migration in Thailand. International Organization for Migration, Regional Office Bangkok, Bangkok

Huijsmans R (2007) Approaches to Lao minors working in Thailand. Juth Pakai (Lao Dev J) 8:18–33

Huijsmans R (2014) Gender, masculinity, and safety in the changing Lao-Thai migration landscape. In: Truong TD, Gasper D, Handmaker J, Bergh SI (eds) Migration, gender and social justice: perspectives on human insecurity. Springer, Heidelberg, pp 333–349. https://doi.org/10.1007/978-3-642-28012-2_19

Huijsmans R, Phouxay K (2008) Whether you go illegally or legally in the end it's the same, you're cheated: a study of formal and informal recruitment practices of Lao workers migrating to Thailand. International Labor Organization, Vientiane

Jayne TS, Chamberlin J, Headey DD (2014) Land pressures, the evolution of farming systems, and development strategies in Africa: a synthesis. Food Policy 48:1–17. https://doi.org/10.1016/j.foodpol.2014.05.014

Jones GW (2013) The population of Southeast Asia. Asia Research Institute Working Paper Series No. 196

Kaur A (2010) Labour migration in Southeast Asia: migration policies, labour exploitation and regulation. J Asia Pac Econ 15:6–19. https://doi.org/10.1080/13547860903488195

Manivong V, Cramb R, Newby J (2014) Rice and remittances: crop intensification versus labour migration in southern Laos. Hum Ecol 42:367–379. https://doi.org/10.1007/s10745-014-9656-6

Martin P (2007) The economic contribution of migrant workers to Thailand: towards policy development. ILO subregional Office for East Asia, Bangkok

Massey DS, Arango J, Hugo G, Kouaouci A, Pellegrino A, Taylor JE (1993) Theories of international migration: a review and appraisal. Popul Dev Rev 19:431–466. https://www.jstor.org/stable/2938462

Molland S (2010) 'The perfect business': human trafficking and Lao-Thai cross-border migration. Dev Chang 41:831–855. 10.1111/j.1467-7660.2010.01665.x

Mori K (2009) Strategy for control of illegal migration to Thailand by the Lao government: case study from the Mekong basin in Laos. Nenpō Tai Kenkyū (J Thai Stud) 9:99–112 (in Japanese)

Nishimoto F (2011) Laos shōsū minzoku no kako 50 nen no jinkō fukugen (Population reconstitution in the past 50 years of ethnic minorities in Laos). Jinkō Gaku Kenkyū (J Popul Stud) 47:126 (in Japanese). https://doi.org/10.24454/jps.47.0_126_1

Niwa T, Nakagawa S, Takahashi S, Nishimoto F (2021) Transnational labor migration from rural village in Laos to Bangkok, Thailand: focusing on relationship with the home village. Nenpō Tai Kenkyū (J Thai Stud) 21:17–30 (in Japanese)

Otomo N (2018) Situation and problems of employment of unskilled foreign workers in Thailand. *Ajia Taiheiyō Tōkyū* (J Asia-Pac Stud) 33:109–143. (in Japanese). https://doi.org/10.57278/wiapstokyu.33.0_109

Pham BN, Hill PS (2008) The role of temporary migration in rural household economic strategy in a transitional period for the economy of Vietnam. Asian Popul Stud 4:57–75. https://doi.org/10.1080/17441730801966683

Pholphirul P, Rukumnuaykit P (2010) Economic contribution of migrant workers to Thailand. Int Migr 48:174–202. https://doi.org/10.1111/j.1468-2435.2009.00553.x

Phouxay K (2017) Patterns and consequences of undocumented migration from Lao PDR to Thailand. In: Bouté V, Pholsena V (eds) Changing lives in Laos: society, politics, and culture in a post-socialist state. NUS Press, Singapore, pp 350–373

Podhisita C (2017) Household dynamics, the capitalist economy, and agricultural change in rural Thailand. Southeast Asian Stud 6:247–273. https://doi.org/10.20495/seas.6.2_247

Qin H (2010) Rural-to-urban labor migration, household livelihoods, and the rural environment in chongqing municipality, southwest China. Hum Ecol 38:675–690. https://doi.org/10.1007/s10745-010-9353-z

Rigg J (2007) Moving lives: migration and livelihoods in the Lao PDR. Popul Space Place 13:163–178. https://doi.org/10.1002/psp.438

Rigg J, Salamanca A, Thompson EC (2016) The puzzle of east and southeast Asia's persistent smallholder. J Rural Stud 43:118–133. https://doi.org/10.1016/j.jrurstud.2015.11.003

Rungmanee S (2016) Unravelling the dynamics of border crossing and rural-to-rural-to-urban mobility in the northeastern Thai-Lao borderlands. Popul Space Place 22:693–704. https://doi.org/10.1002/psp.1989

Sciortino R, Punpuing S (2009) International migration in Thailand 2009. International Organization for Migration, Thailand Office, Bangkok

Skeldon R (2006) Recent trends in migration in east and southeast Asia. Asian Pac Migr J 15:277–293. https://doi.org/10.1177/011719680601500206

Southichack M (2013) Lao labor migration and remittance: trends and economic and livelihood implications. Cooperation (Laos Program), Vientiane

Thomson SS, Baden S (1993) Women and development in Laos. Institute of Development Studies, University of Sussex, Brigton

Thongyou M, Ayuwat D (2005) Social network of Lao transnational migrant workers in Thailand. Research Centre (SEARC) of the City University of Hong Kong

Tomita S, Parker DM, Jennings JA, Wood J (2015) Household demography and early childhood mortality in a rice-farming village in Northern Laos. PLoS ONE 10:e0119191. https://doi.org/10.1371/journal.pone.0119191

Growth in the Use of Modern Contraceptive Methods and the Related Social Context in a Central Laotian Farming Village

Chihiro Shirakawa, Futoshi Nishimoto, Shinichi Takahashi, and Satoshi Yokoyama

Abstract The purpose of this chapter is to reveal the process and characteristics of the growth in the use of modern contraceptive methods in a central Laotian farming village, as well as to consider the social context and background against which that spread took place. In the village that is the subject of this study, the birth rate has fallen since the second half of the 1990s, and the average number of births per woman has been declining as well. It is essential to examine the spread of the use of modern contraceptive methods when seeking to understand such changes. It has become evident that most women in the studied village use such methods, the use of which spread starting in the second half of the 1990s. Meanwhile, in understanding the societal context that led to this spread, a key point is the Lao government's family planning program that began in 1996. Many women may have begun to use contraceptive methods because of this program. However, a considerable number of women had started using it before this point, the trigger for which can be identified as working abroad in Thailand.

Keywords Modern contraceptive use · Women · Government family planning program · Working abroad · Laos

---

C. Shirakawa (✉)
Graduate School of Human Sciences, Osaka University, Suita, Japan
e-mail: itakuma@hus.osaka-u.ac.jp

F. Nishimoto
Embassy of Japan in Laos, Vientiane, Laos
e-mail: nsmtfts@gmail.com

S. Takahashi
Kobe University, Kobe, Japan
e-mail: takahashi@econ.kobe-u.ac.jp

S. Yokoyama
Graduate School of Environmental Studies, Nagoya University, Nagoya, Japan
e-mail: yokoyama.satoshi.i4@f.mail.nagoya-u.ac.jp

S. Yokoyama et al. (eds.), *Population Dynamics and Livelihood Changes of Small-Scale Societies in Laos*, International Perspectives in Geography 22,
https://doi.org/10.1007/978-981-96-2956-5_6

1 Introduction

The phenomenon of declining birth rates has been observed for some time in Japan, Western countries, and South Korea. However, in recent years, it has also become apparent in other countries, including Southeast Asian countries such as Thailand and Laos. The causes of this phenomenon have been identified, including rising levels of education and income, declining mortality rates, and urbanization (Kono 2007: 118–129; Otsuka 2015: 231–232). However, looking at the micro level of smaller regions and communities, rather than the macro level of nations, the question of how exactly this phenomenon has advanced has not yet been fully clarified.

Previous studies have frequently identified, in addition to the above-mentioned list of causes of declining birth rates, the ease of procuring family planning methods, which are directly related to birth. For example, Aoyama et al. (2001: 127) state that "lower child mortality, improved education for women, and access to family planning methods have always been recognized as affecting lower birth rates." Based on these indications, as a first step toward revealing the micro process of declining birth rates, this chapter aims to clarify the process and characteristics of growth in the use of modern contraceptive methods among people. It also aims to discuss the social background that led to the growth, in a farming village in central Laos (hereinafter referred to as "KH village").

Social science studies on the use of modern contraceptive methods in Laos can be broadly classified into two categories. The first category focuses on the actors carrying out family planning policies or the actors offering knowledge and supplies (e.g. Sychareun 2004; Sychareun et al. 2010; Shimazawa 2014. The second category examines the actors receiving and utilizing the resources provided by other actors. This chapter falls into the second category, and as a study belonging to this group of research, we can start by listing investigative reports by governments and international organizations that have investigated the facts regarding the kinds of knowledge people have about family planning, and which types of contraceptive methods they use (e.g. Ministry of Health and Lao Statistics Bureau 2012: 85–94; UNFPA 2015; Lao Statistics Bureau 2018: 90–98). Their purpose is to understand the overall situation in Laos, with a focus on the macro level.

Meanwhile, while the number of such studies are limited, some studies have been conducted at the micro level, such as the village studied in this chapter (e.g. Sychareun et al. 2013; Shimazawa 2014; Chantakoumane et al. 2020). Such studies have placed importance entirely on ascertaining user trends with regard to contraceptive methods at a particular point in time or analyzing the causes that dictate these trends. However, when attempting to clarify the process by which the use of contraceptive methods spread as in this chapter, it is essential to grasp the user trends and the social context surrounding users and to take a longitudinal approach to a certain extent. Then, it is necessary to consider the correlation between the user trends and the social context surrounding users, and to determine whether they have any kind of causal relationship over time. Nevertheless, few studies have been conducted on the basis of such a point of view and approach.

A comparable trend has been observed in studies conducted in Thailand, which shares a similar cultural background to Laos. For example, in the Nang Rong District, Buriram Province of northeastern Thailand, which is relatively close to central Laos, where KH village is located, the University of North Carolina at Chapel Hill in the United States and the Mahidol University in Thailand had collaborated on a joint research project.[1] The study conducted three large-scale social surveys from 1984 to 2000 targeting approximately 34,000 people in 51 villages within the district. The results revealed changes in demographic and socioeconomic conditions, as well as the use of contraceptive methods (e.g. Entwisle et al. 1996; Edmeades 2008). However, there has been insufficient historical examination of the correlations or causal relationships between these changes and changes in each village's socioeconomic conditions. For this reason, this chapter seeks to fulfill the above-mentioned purpose while attempting to compensate for the above aspects that were found to be lacking in previous studies.

The research for this chapter was conducted in KH village, located about 65 km southeast of the Laotian city of Savannakhet. The village is administratively part of Songkhone District, Savannakhet Province. As of the end of 2014, the population was 760, with 112 households. All villagers are Lao, the majority ethnic group in Laos.[2] The primary occupation of the people is rice agriculture using rain-fed paddy fields surrounding the village. However, it is also common for people to engage in transnational labor migration to Thailand (Niwa et al. 2021: 20–23).

The birth rate has fallen in Laos from the 1990s onward, and KH village is no exception. According to a survey by Takahashi et al. (2023: 268–270), the birth rate in KH village had hovered around 40 births per 1,000 people until the early 1990s. However, it dropped precipitously to around 15 births per 1,000 people starting in the late 1990s. As a result, while the village population maintained a growth rate of about 2.5% per year from the 1950s to the late 1990s, the growth trend slowed down after that. Recently, growth rate has exhibited a slight decline.

The average number of births per woman in KH village has also tended to decline, from 5.7 per woman aged 45 and older, to 3.2 per woman aged 35 to 44 (Takahashi et al. 2023: 269). It should be noted that women aged 44 and under still have the possibility of having more children in the future. However, when considering the ages at birth of these women aged 45 and over, it seems unlikely that women in each age group below 44 would surpass an average of five births. In order to understand the trends surrounding births in KH village, it is essential to consider the spread of modern contraceptive methods.

This chapter presents the results of a research project headed by coauthor Satoshi Yokoyama.[3] In this project, coauthor Shinichi Takahashi and colleagues conducted

[1] For details on this social survey, called the Nang Rong Project, see the following website (https://www.cpc.unc.edu/projects/nangrong/index.html, last accessed October 5th, 2019).

[2] In this chapter, we use the term Lao to refer to the ethnic group, and the word Laotian to indicate citizens of Laos, which includes Lao and other ethnic groups.

[3] This project was run through KAKENHI Grant-in-aid for Scientific Research (A), Population Dynamics, Reproduction and Livelihood Changes in Small-scale Communities in Laos (JP25257004, Project Period: 2013–2016).

surveys of all households in December 2013, September 2014, and December 2014. The surveys were primarily conducted by asking questions orally using a questionnaire sheet. The questionnaire included questions on each household's family composition, occupation, and ownership of fields and other property. It also includes several questions on their use of modern contraceptive methods.

Furthermore, this chapter's lead author Chihiro Shirakawa conducted informal interviews with a narrower focus on the use of modern contraceptive methods, for 45 women from KH village. He did so with the cooperation of the chapter's coauthor Futoshi Nishimoto, Ms. S (a female registered nurse serving at a health center in LN village, neighboring KH village), and female staff at the Savannakhet Provincial Health Department who provided support as interpreters, etc.[4]

Additionally, he conducted informal interviews with other key informants, including medical product vendors and medical professionals, to understand medical product distribution and trends in public policy related to medical care in KH village. These interviews were conducted in December 2014, November 2015, and September 2016. This chapter is based on data and knowledge gained from the above-mentioned series of surveys and interviews.

2 The Status of Modern Contraceptive Use

The all-household questionnaire-based survey conducted in December 2013 revealed that a total of 109 people in KH village had experience using some kind of modern contraceptive method. All of them were women. In contrast, no men in the village were found to regularly use condoms or had undergone a vasectomy or other contraceptive surgery. This indicates that in KH village only women use contraceptive methods.

The reasons behind the men's reluctance to use condoms or have contraceptive surgery remain unclear. The results of our survey did not yield sufficient information to clarify this phenomenon, and the pertinent details remain elusive. Shimazawa (2014: 214) lists reasons for this in other parts of Laos, including that "the widespread perception is that women are the ones that do family planning," as well as male-dominated gender relations.[5] A married woman in her thirties in KH village said, "I never know when my husband will ask for it. Sometimes he wakes me up when I'm

[4] In selecting participants for informal interviews, we primarily focused on the 30 to 49 age bracket. Note that while the interview participants were all women, the all-household survey questionnaire was sometimes responded to by a male head of household, etc. Consequently, the interview participants and the all-household survey respondents are not entirely overlapping groups, and for this reason, the interview and all-household survey data content presented in subsequent sections naturally differ.

[5] However, regarding the fact that contraceptive methods were used exclusively by women among the lowland Laotians, including Lao, in some cases maternal kinship theories have been highlighted as a possible reason for women's autonomy from men surrounding sexuality and bearing children (Schenk-Sandbergen 2012: 84).

sleeping. It's always at his pace. He doesn't say 'Let's do it' beforehand. So the only thing I can do is take the pill." Given this account, it is plausible that Shimazawa's highlighted reasons may also apply to the people of KH village.

In the all-household survey, we also asked about specific methods among 109 women that responded as having had experience using modern contraceptive methods (multiple answers allowed). According to their responses, 66 had experience using oral contraceptives (the pill), 46 had used contraceptive hormonal injections, seven had had tubal ligation surgery, and two had used an intrauterine contraceptive device (coil). Furthermore, 21 of them had used multiple methods, with 19 having used the pill and injections.[6]

As previously mentioned, informal interviews focused on the use of modern contraceptive methods were conducted. Of the 45 women interviewed, 34 had experience using some kind of contraceptive method. Looking at specific methods in order of most common, 24 having used injections, 15 having used the pill, two having used the coil, and one having undergone tubal ligation surgery (multiple answers allowed, Table 1). Furthermore, eight of them had used multiple methods, with five having used the pill and injections.

The results of the two surveys indicated that the pill and injections were common among women in KH village.[7] Of these, the pill was available for purchase at two of the four small stores in the village. Both of these stores procured their Thai products from Thai city of Mukdahan on the other side of the Mekong River from Savannakhet. Alternatively, they obtained the same products from a pharmacy in Paksong, the town closest to KH village where the Songkhone District office is located. The price for a sheet of 28 pills (for 28 days) was 4,000 kip at one store and 5,000 kip at the other store.[8]

Aside from these two small stores, the pill could be procured from a woman in KH village who had experience working as a nurse. Aged 64 (as of 2014), this woman was from Houaphanh Province in northeastern Laos. After working there as a nurse, she married her husband who was a former soldier from KH village and had been living in KH village since 1996. She procured a different type of Thai-manufactured

[6] The remaining responses were one each for the pill and the coil, and injections and the coil.

[7] While Tomita and Badenoch (2014: 266) do not show quantitative data, they state that many women had had tubal ligation surgery in a farming village in the northern Laotian province of Oudomxay, their area of study. Compared to this, while the main cause is unclear, there were few such cases in KH village, as seen above. According to Matsuo (2013: 76–84), tubal ligation surgery began to spread in farming village regions of the western Indian state of Maharashtra in the 1970s, and by the mid-2000s female sterilization procedures, including tubal ligation, accounted for more than 90% of modern contraceptive methods used by the people there. The main reason for this was that as it became commonplace for women to give birth at hospitals and other medical facilities, healthcare professionals would present childbirth followed by sterilization surgery as one course of events for expectant mothers, as a part of maternal and child health activities. We did not observe a similar trend surrounding sterilization procedures among the women of KH village, but there were more than a few examples of medical professionals recommending the use of the pill or injections following birth at a facility, including Case 2 shown in Sect. 6.

[8] The exchange rate of the local currency the kip to the Japanese yen was 1.5 yen to 100 kip as of December 2014.

Table 1 Contraceptive use of the interviewees of the informal interviews

No	Age in 2014	Used contraceptive measures	Starting year of contraceptive use
1	55	Injection	1996 (after the birth of the 5th child)
2	54	Injection	2000 (after the birth of the 7th child)
3	51	Injection	Early 1990s? (after the birth of the 2nd child)
4	48	Pill → Injection	1996 (after the birth of 7th child)
5	46	Injection	Not available
6	45	Injection → Pill	1994 (after the birth of the 1st child)
7	44	Injection	1999 (after the birth of the 3rd child)
8	44	Injection	1992 (after the birth of the 2nd child)
9	43	Pill	2003 (after the birth of the 3rd child)
10	43	Injection	1999 (after the birth of the 4th child)
11	43	Injection	2004 (after the birth of the 4th child)
12	42	Injection → Tubal ligation	1993 (after the birth of the 4th child)
13	41	Pill → Injection → Pill	1990 (after the marriage)
14	41	Injection	2004 (after the birth of the 5th child)
15	40	Injection	2013 (after the birth of the 3rd child)
16	39	Pill	1999 (after the birth of the 1st child)
17	39	Injection	2005 (after the birth of the 2nd child)
18	39	Injection	2005 (after the birth of the 3rd child)
19	38	Injection	2000 (after the birth of the 2nd child)
20	36	Injection	2004 (after the birth of the 5th child)
21	35	Injection → Coil	1999 (after the birth of the 1st child)
22	33	Pill	Not available
23	32	Pill	2002 (after the marriage)
24	32	Injection	2011 (after the birth of the 1st child)
25	31	Pill	1998 (after the marriage)
26	29	Pill → Injection	2007 (after the marriage)
27	29	Injection → Coil → Injection	2003 (after the birth of the 1st child)
28	28	Pill	2010 (after the birth of the 3rd child)
29	27	Pill	2006 (after the marriage)
30	26	Injection	2004? (after the birth of the 1st child)
31	26	Pill	2004 (after the marriage)
32	21	Pill	2013 (after the birth of the 1st child)
33	21	Pill	2013 (after the birth of the 1st child)
34	20	Injection → Pill	2012 (after the birth of the 1st child)
35	78	Unused	
36	73	Unused	

(continued)

Table 1 (continued)

No	Age in 2014	Used contraceptive measures	Starting year of contraceptive use
37	64	Unused	
38	55	Unused	
39	53	Unused	
40	52	Unused	
41	48	Unused	
42	45	Unused	
43	44	Unused	
44	42	Unused	
45	22	Unused	

Source Author's informal interviews

pill than the two stores were offering at a pharmacy in Paksong. These were being sold for 7,000 kip per sheet.

Meanwhile, regarding injections, the pharmaceuticals and syringes used for these were not sold in KH village. Instead, the above woman from Houaphanh Province was giving the injections to those who were interested in 11,000 kip per shot. The pharmaceutical used was Belgian-made Depo Provera, which she procured from a pharmacy in Paksong. The effects of a single injection continued for three months.

In KH village, there was another person who was giving a different kind of injection. Similar to the above woman from Houaphanh Province, there was a woman in a neighboring village to KH village who was 50 years old (as of 2014) and had previously worked as a nurse. Her husband was from KH village and she had used to live there. At of the time of the survey, the woman and her husband were living in her home village. She made multiple visits to KH village on a motorbike each week to sell medicines and daily necessities. While she did not offer the pill, she provided injections for 10,000 kip each. The pharmaceutical used was the same as that of the woman from Houaphanh Province, Depo Provera procured at a pharmacy in Paksong.

3 Source of Contraceptive Method Information and History of Use

In the all-household survey, we also asked the 109 people who had experience using modern contraceptive methods about their information sources and their history of use. Regarding the former question, we received responses from 66 people. The breakdown was as follows: 40 people (nearly 60%) received information from healthcare professionals or medical institutions, 16 people received information from friends or family, and 10 people received information from Bangkok, Thailand.

The results showed when the information about contraceptive methods to women of KH village was facilitated by the involvement of healthcare professionals and medical institutions. Of those, the most common answer was nurses (19 people). However, the location of employment of the nurses could not be determined. Although there are no medical facilities in KH village, LN village offers a health center located about 90-min walk away, and the Songkhone District Hospital in the town of Paksong located about a two-hour walk away. The nurses listed as sources of information may have been nurses at these medical facilities. After nurses, the most common answers were the district hospital (six answers), "hospital in Paksong" and "hospital" (four each). Since it is highly likely that "hospital in Paksong" and "hospital" were referring to the district hospital, these three answers collectively account for 14 people.

Of the 16 people whose source of information on contraceptive methods was friends or family, ten listed friends as the most common answer. This was followed by mother (four people) and sister (two people).

What is noteworthy is the response of Bangkok, from ten people. As previously mentioned in Sect. 1, going abroad to Thailand for work is common in KH village, and these ten people received information about contraceptive methods in Bangkok, where they went to work. In order of most common responses, the breakdown is friends (four people), healthcare professionals or medical facilities (three people), employer (two people), and television (one person).

It is also noteworthy that some of these ten people started began using contraceptive methods an early stage. For example, one of the women who listed Thai healthcare professionals or medical facilities as her information source began using it in 1990 after receiving information from a doctor in Bangkok.[9] Furthermore, the two people who listed their employer as their information source started using it in 1991.[10]

Meanwhile, 94 people responded to our inquiry regarding the history of their use of contraceptive methods. The most common response was that they started using the method after it was recommended to them by a doctor, with 50 people giving this response (just over half of all respondents). The next most common response, with 30 people, was that they were recommended the method by a friend. These two responses alone account for over 80% of all responses. The most common remaining was "I used it because many other people were using it" (four responses).

Based on the above points, it can be inferred that healthcare professionals and medical facilities played a significant role in the dissemination of information regarding contraceptive methods and decision-making process in their use among women in KH village. However, based on the above series of responses about their histories of using contraceptive methods, the reasons remain unclear as to why the respondents followed the recommendations of doctors and friends and came to use these methods. Thus, we attempted to ask about this in the informal interviews.

---

[9] This women's case is taken up shortly in this section, as Case 1.

[10] One was born in 1966, the other in 1970.

As mentioned in the previous section, of the 45 women we interviewed, 34 responded as having experience using contraceptive methods. We were also able to ask half of them, 17 people, the reasons that led them to use it. Seven of them cited the difficulty of childbirth and childcare. The next most common response, cited by five people, was the economic burden of childcare. Respondents indicated that they had used contraception to try to avoid having children because childrearing required many expenditures and the family was struggling financially, which fall into this category. Another reason given by four people was that they wanted to plan their childbirths, for example, to ensure enough time between each birth.

The last person was the above-mentioned woman who received information about contraceptive methods from a doctor in Bangkok and began using them in 1990 (No. 13 in Table 1 in the previous section). As the following case indicates, her reason for using contraceptive methods is related to going abroad to Thailand for work.

<Case 1>

Person A, born in 1973, went to work in Thailand for the first time when she was 26 (in 1989) and began working at a garment factory in Bangkok. She married the following year and they lived together while both worked. However, she was afraid that if she got pregnant she would be dismissed from her job and sent back to Laos, so she learned about contraceptive methods from a doctor, and started taking the pill when she started living with her husband (in 1990). A Thai coworker told her that the pill was safer than the injections, and all of her female married Laotian coworkers were using the pill.

In understanding the processes by which the use of contraceptive methods spread among the women of KH village, we cannot ignore leaving to work in Thailand, as seen in this case. We will examine this point in detail in the following sections.

4 Period in Which Respondents Started to Use Contraception

In the all-household survey, we asked the 109 women who had experience using contraceptive methods about the period in which they started using. We received 100 responses to this question. The results are summarized in Fig. 1. This shows that the number of people who started using increased conspicuously in the late 1990s, in particular around 1997 seems to be a turning point. As we touched on in Sect. 1, the late 1990s is also when the birth rate in KH village dropped abruptly, and from this, we can see that one of the factors in the falling birth rate in KH village is the spread of the use of modern contraceptive methods.[11]

[11] Tomita and Badenoch (2014: 266) list the spread of modern contraceptive method use as one factor in the decline of the birth rate starting in 1989 in a farming village in Oudomxay Province, northern Laos. In this village located close to the border with China, the pill began to be offered in 1999, and movement to the border became easier with infrastructure improvement. As a result, more women underwent tubal ligation surgeries at Chinese hospitals starting in the 2000s.

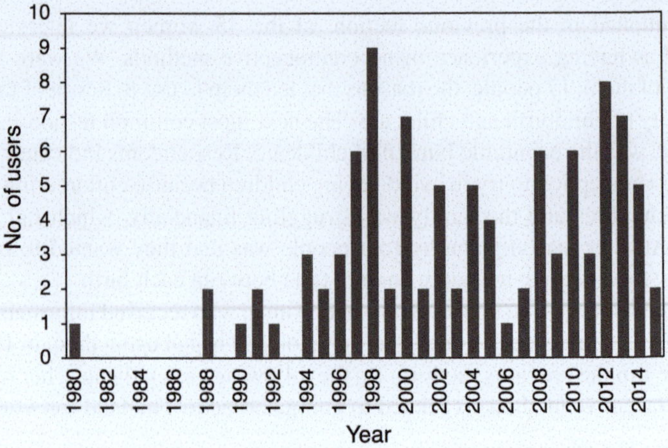

Fig. 1 Starting year of contraceptive use. *Source* Authors' household survey

Meanwhile, as mentioned in Sect. 2, we conducted informal interviews with 45 women, 34 of whom had experience using some kind of contraceptive method (Sect. 2, up to No. 34 in Table 1). Among these women, of the 15 women aged 40 and older (No. 1 to No. 15 in Table 1), 11 of them began using it after giving birth multiple times, ten of whom began using it after having three or more children. Furthermore, seven of these 11 women began using it in the 1990s.

In contrast, of the 19 women in their 30 s or younger (No. 16 to No. 34 in Table 1), 13 of them began using after the birth of their first child or immediately after marriage. Furthermore, ten of these women began using in the 2000s or later.

In light of the above clarifications, among the women of KH village, contraceptive methods were first used among women who had been pregnant multiple times and had multiple children, as a way of not having any more children. Then, afterward, it is conceivable that among the younger generation of women, contraception was used as a way to control births in a planned manner starting in the early stages after marriage.

Based on the above clarifications, we can ask at least the following two questions related to the purpose of this chapter, as stated in Sect. 1. One is the question of what happened around 1997, the turning point in the use of contraceptive methods. The other is how the 12 women who started using in 1996 or earlier came to use, since this is no small number (Fig. 1). We will examine the former question in the next section, and the latter question in Sects. 6 and 7.

5 The Lao Government Family Planning Program

Concerning the first question, it is worth noting the Lao government's family planning program. According to Shimazawa (2007: 157–158; 2014: 195–204), the Lao government's family planning policy began to be implemented in the 1970s, but was not fully implemented until the 1990s. The goal was not population control, but rather to adjust birth spacing to improve women's health. For this reason, activities related to family planning were run as a part of maternal and child health programs. Agencies that played a central role in implementation included the Lao Women's Union[12] (*Sahaphan Mae Ying Lao*) and countries providing official development assistance to Laos in the early 1990s, and the Ministry of Health and the United Nations Population Fund (UNFPA) in the late 1990s. Initially, these programs were limited to urban areas, but in the late 1990s they started to expand to rural areas.

In 1996, the government's family planning program started to be implemented in Songkhone District, where KH village is located. For example, the district hospital in Paksong held four to five workshops per year on different kinds of topics related to "development and women." These workshops, which lasted from two to four days, included family planning and lectures on all kinds of contraceptive methods. Participants were also informed that contraceptives for these methods were readily available at the district hospital, that the pill and condoms were free, and that injections were also free.

Women's Union members from each village in Songkhone District attended these workshops, and after the workshops ended, the women returned to their respective villages and held meetings with the cooperation of the village head. They then shared the information they gained from the workshops with the villagers. In the case of KH village, the meetings were held at a Buddhist temple in the village, and the village head asked that one person from each household attend. Although the meetings were held in a top-down fashion, most of the villagers who attended showed interest in receiving information about contraceptive methods.

As mentioned in the previous section, the use of contraceptive methods spread widely among the women of KH village starting in the late 1990s. Above all, the number of people who started using in the year after 1996, when the government family planning program began, doubled relative to the previous year (Fig. 1 in the previous section). Keeping in mind the consistency of this timing, we can see that the start of the government family planning program was a key factor that led to the turning point at which women started to use contraceptive methods around 1997.

Note that we were unable to confirm the implementation period of the program.[13] However, official family planning activities in KH village were ongoing, albeit on an irregular basis, at least from 2006 until the time of the survey, depending on the conditions of budget availability. These activities were led by female Nurse S at the LN village health center and comprised five activities related to maternal and child

---

[12] For details on the Lao Women's Union, see the work of Iinuma (2013).

[13] According to the district hospital nurse, at the time the program began in 1996, free provision of the pill and injections was conducted with the support of UNICEF and continued until 2005.

health; perinatal health screenings for pregnant women; physical measurements of children up to five years of age; vaccinations; family planning; and hygiene education. Regarding family planning, specific information about contraceptive methods was provided, and injections were given to those who wanted them. However, in 2014 these activities were held only twice, in June and October, due to budget constraints. Furthermore, family planning activities were planned alongside the other four activities at this time, but most of the women already had knowledge of contraceptive methods and were already receiving injections at other places, which may have been why implementation was abandoned due to lack of participants.

6 Working Abroad in Thailand and the Use of Contraceptive Methods

In the previous section, we noted that a main factor in the turning point at which women in KH village started to use modern contraceptive methods around 1997 was the start of the Laotian government's family planning program. However, Fig. 1 shows that a considerable number of women began using contraceptive methods before 1996, the year in which the program started. This raises questions relating to the circumstances that led to women starting to use contraceptive methods at that time. When we looked into the common features of these women to answer this question, the experience of working abroad in Thailand featured in many responses.

People from KH village were already leaving to work in Thailand in the 1980s, but at that time the number of workers was still small. However, in the all-household survey of December 2013, 177 people out of a population of 629 had gone to work in Thailand. It was revealed that the percentage of all households that had someone working away from home had risen to 73%. Most migrant workers were in relatively younger age brackets, with 78% of all migrant workers being men and women aged 34 and under. In the 20–34 age bracket, roughly 60% of men and women in this bracket had left home to work abroad. Their destination was the Thai capital of Bangkok and its surrounding area, and the general form of work was unskilled labor with irregular employment in places such as restaurants and urban factories.[14]

Although this was a young age group, those under the age of 34, who comprised nearly 80% of the migrant workers, were not always unmarried. There were, of course, some with spouses and children. When these people left KH village for work, it was common for them to leave their children with grandparents or other relatives in KH village and go to Thailand as a couple, with both of them working.[15] The income would inevitably increase if both of them worked instead of just one person in the

[14] Niwa et al. (2021) gives detailed descriptions and analysis regarding the circumstances surrounding the people of KH village going abroad to Thailand to work.

[15] Kiso (2013: 58–65; 2016: 99–105) reports that in rural Maha Sarakham Province in northeastern Thailand, it was once the case that only unmarried women were observed going to work in Bangkok and the surrounding areas, but starting in the late 1980s, it became more widespread for married women with young children to do so as well. Kiso also notes that migrant work among married

couple. However, if a couple has children, especially young children, and brings them to Thailand, the wife would be compelled to devote herself to childrearing in Thailand because they would have no parents or relatives there. The burden of childrearing would increase while the income would decrease. Furthermore, families with children may find it impossible to live in Thailand on the husband's income alone. Because of these problems, people would leave their children in KH village and both husband and wife would go to work in Thailand.

However, even if they left their children behind in KH village and both husband and wife left to work, if they had another child while at their work destination, they would be confronted with the exact problem they had tried to avoid, making it meaningless to have gone to the trouble of leaving their other children in the village. For this reason, the wife would use contraception as birth control at the work destination. The following is an example of such a case, along with a brief biography of the woman involved.[16]

<Case 2>

Person B was born in 1971 (Sect. 2, No. 9 in Table 1), married at age 19 (in 1990), had her first daughter the following year and her second daughter when she was 23 (in 1993). She went to Thailand to work for the first time when she was 25 (in 1996). She obtained her passport and a Thai visa, departed the country from Savannakhet to Mukdahan in Thailand, and took a public bus to Bangkok. Her reason for working away from home was to earn money for raising her children. In Bangkok, she worked at a car wash for two years, and then briefly returned to Laos. One year later, she returned to Bangkok a second time and worked at construction sites. Her daily wage was 160 baht.[17] We were unable to confirm whether she went with her husband the first time she worked away from home, but the second time she did go with him, and they left their daughters in the care of her parents in KH village.

As Person B got older, she thought she could no longer get pregnant, so she stopped using contraception. However, when she was 33 (in 2003), she found out that she was pregnant. Because of this, she returned to Laos in the third month of her pregnancy and gave birth to her third child (her first son). Three months after giving birth, she started to take the pill on the recommendation of a doctor at the Paksong District Hospital, where she had undergone her prenatal examinations. This doctor was involved with maternal and child health activities at the district hospital.

When her son was 13 months old (in 2004, when Person B was 34), she left her children with her parents and once again went back to work in Bangkok with her husband. She worked for one year, returned to childcare, and then returned to Bangkok for another year later. Since it would have been difficult for her to get pregnant while in Bangkok, she took the pill while working there. When she was 39

women was facilitated by the presence of their mothers, aunts and other close female relatives who raised the children in their place, and this also applies to KH village.

[16] This life story was recounted in an informal interview. The same is true for all the cases from Case 3 onward, as shown in the following sections.

[17] The exchange rate of the Thai baht to the Japanese yen was about 1 baht to 4 yen in the mid-1990s, about 3 yen after 2000, and hovering at about 3.5 yen since the late 2010s.

(in 2010), she returned to Laos and had her fourth child (third daughter) the following year, after she did not migrate again for work.

As with Person B, it was common among the women from KH village who went to work in Thailand with their husbands to use some kind of contraceptive method while they were away working. Some women did not use contraception, and again, like Person B, they had no expectation of becoming pregnant and thought it was impossible because of their age. Consequently, when an "unforeseen circumstance" arose, many of them went back to Laos to give birth and raise children, just like Person B.

While she was away working, Person B used contraceptive methods of her own volition, to avoid problems that would arise if she became pregnant. However, there were also cases like that of Person A, Case 1, shown in Sect. 3, in which people used contraception whether they liked it or not because they were afraid that if they became pregnant, their employer would fire them and send them back to Laos. The present study found only one case like that of Person A. Since the married Laotian women who were Person A's colleagues while working away from home all used the pill, there may have been more such cases.

7 Working Abroad and Constraints on the Use of Contraceptive Methods

In any event, most of the women used modern contraceptive methods to avoid the problems associated with becoming pregnant while working away in Thailand. However, for some women, it was not always easy to use contraceptive methods, as indicated by the following case.

<Case 3>

Person C (Sect. 2, No. 16 in Table 1) was born in 1975 and went to work on a lemon plantation in Prachuap Kiri Kahn Province, located to the southwest of Bangkok, when she was 16 (in 1991). After working for three years for a monthly salary of 100 baht, she moved to Bangkok, and worked for two years in a confectionery factory. This time, her monthly wage was 200 baht.

When she was 23 (in 1998), she gave birth to her first son while she was working in Bangkok. She got married at that time, and she and her husband returned to Bangkok for work. She wanted to return to Laos to give birth, but was unable to for the lack of money.

Person C learned about contraceptive methods from a friend from KH village who was working with her in Bangkok. According to Person C, contraception such as the pill was not sold at local pharmacies in Bangkok at that time. She could nonetheless buy the pill at hospitals, which also offered injections. However, Person C did not have a passport or visa, and could not go to the hospital for fear of being asked to show her documents. If she had not had the fear of being discovered as residing there illegally, or if she had had a passport, she said she would have received injections at the hospital from the start, because it would have been difficult to raise children while working in Bangkok. Ultimately, after giving birth, Person C devoted herself to childcare, and only her husband worked.

Person C first used a contraceptive method when she was 24 (in 1999). While on a temporary trip back to Laos, she purchased the pill at a pharmacy in Paksong and started taking it. The following year, however, she found out she was pregnant with her second child while working in Bangkok. Again, she did not have a passport or visa this time either, but fortunately she had the money to return to Laos. So instead of going to a hospital in Bangkok, she went back to Laos and gave birth to a boy.

Person C did know contraceptive methods before her first son was born in Bangkok, and if circumstances had allowed, she would have used the pill or another method. However, if she had gone to the hospital to get the pill or injections, there was a possibility that she would have been asked to show her passport, and she could have been found to have entered and stayed in Thailand without obtaining these documents.[18] Because of her fears and concerns about this kind of situation arising, she was unable to use contraceptive methods even if she wanted to.

Considering the above situation, even though Person C gave birth in Bangkok, she may not have done so in a hospital. We were unable to confirm this point with her, but in truth, as we will see in the two cases reported below, multiple people had given birth without using hospitals. Nonetheless, as those two cases will show, some medical facilities, such as private clinics, did not require patients to show a passport, and in a small number of cases, people obtained the pill or other forms of contraception from such places.

<Case 4>

Born in 1970, Person D (No. 8 in Table 1) gave birth to her first daughter when she was 16 (in 1986) and to her first son when she was 18 (in 1988). Then, the year after the birth of her son (in 1989, when she was 19), she and her husband went to work in Thailand. They worked in the same initial workplace as Person C in Case 3, the lemon plantation in Prachuap Kiri Kahn Province.[19] They left their children with her parents in KH village. Neither had passports or visas.

When she was 22 (in 1992), a doctor at a hospital in Thailand recommended contraceptive methods to her because if she became pregnant it might be difficult to raise children while working, so she received injections. Until that point, Person D did not know modern contraceptive methods. She went to the hospital on instructions from her employer. She was sent to undergo a health examination because if she became pregnant, childcare would make it impossible for her to work, and that would be a problem. Since Person D did not have a passport or visa, she was afraid that her status would be found out if she went to the hospital, so at first, she did not want to go. However, her employer pushed her to go, saying not to worry, and that if anything happened, they would take care of it, so she decided to go.

---

[18] On this point, Phouxay (2017: 368) mentions cases in which not only were Laotian migrant workers found out as having illegal status after visiting the hospital, but they were also arrested by police.

[19] We were unable to confirm whether this was the same plantation at which Person C went to work.

This is how Person D began using contraception, but the following year (in 1993, when she was 23) she became pregnant with her third child. She had forgotten to get her injection. Although she had used the hospital the previous year, she was still deeply fearful of having her illegal residence status found out by going to the hospital. Thus, she gave birth to a girl in a hut at the plantation, without using a hospital. Person D then stayed on the plantation with her husband and child, returning to Laos when she was 30 (in 2000). They set aside the money they earned working away from home to build a new house and purchase a paddy field.

<Case 5>

Person E (No. 6 in Table 1) was born in 1969 and went to work in Thailand for the first time when she was 15 (in 1984). She had seen other people from KH village go earn money in Thailand and decided she also wanted to go. Person E's older sister was working on a plantation near the town of Khemarat in Ubon Ratchathani Province in northeastern Thailand. Person E pretended to her parents that she was going to school with her friend, and went in secret to stay with her sister. She did not have a passport or visa. Several days later, Person E's parents learned her whereabouts from her sister, and they were furious. She was also reproached by her sister for having come to Thailand without telling their parents. However, Person E did not return to KH village but instead left Khemarat for Bangkok, where she worked at a factory manufacturing extension cords near Chinatown. Her monthly salary was 300 baht. Afterward, she returned briefly to Laos, then went back to Bangkok, this time working at a car wash near the royal palace. This time, her monthly wage was 600 baht.

When she was 25 (in 1994), she met and married her husband in Bangkok. She had learned about modern contraceptive methods in Bangkok from her Laotian friends, but she did not use them due to her desire for children. Before long, she became pregnant and took the opportunity to move to the lemon plantation in Prachuap Kiri Kahn Province. She went there because many of her Laotian friends were working at the plantation, and she thought they would help her with childbirth and childrearing.[20] Person E also did not have a passport, and, similar to Person D in Case 4, she gave birth to her first son in a hut at the plantation. Afterward, she devoted herself to childcare until going back to Laos, with only her husband working on the plantation.

After giving birth, Person E went to a private clinic to receive contraceptive injections. She was able to use this clinic because it did not ask patients to show their passports or other documentation. However, perhaps because of side effects from the injections, her physical health worsened and she switched to the pill. Nonetheless, when she was 27 (in 1996), she became pregnant with her second child. According to Person E, this was probably due to irregular use of the pill. Her husband's employer told her not to worry about not having a passport, and they took her to the above hospital, where she gave birth to her first daughter. After giving birth, the doctor prescribed her the pill, and she took it.

---

[20] We were unable to confirm whether this was the same plantation at which Person C from Case 3 and Person D from Case 4 went to work. However, considering that many of her Laotian friends were working there, it may have been the same as at least one of the plantations.

Person E returned to Laos at the age of 31 (in 2000). Due to the difficulties of living in Thailand on her husband's earnings alone, and concerns about continuing to stay illegally, she returned to Laos with her children. She also had the option of sending her children back to Laos without her and continuing to work abroad along with her husband. However, her father had already passed away, and she could not place the children in the care of her aging mother. After that point, Person E did not go back to Thailand for work.

As evidenced by the above two cases presented, both Person D and Person E entered Thailand without passports or visas and continued to work while staying illegally. Person D first went abroad to work in 1989, while Person E did so in 1984. Person C from Case 3 was also residing illegally when she gave birth to her first son in Bangkok in 1998. While this is unconfirmed, considering that she did not have a passport at that time, likely, Person C did not possess a passport during the five years she worked there earlier, starting in 1991.

As we can surmise from these cases, particularly up to the early 1990s, it was not unusual for migrant workers to cross the Mekong River by boat, and enter Thailand in secret from places like Khemarat (as seen in Case 5) without having obtained a passport or visa. Proceeding from KH village and traversing Paksong, one arrive at the Mekong River by following a nearly direct southward course. Khemarat is on the opposite bank, much closer to KH village than other Thai border towns like Mukdahan.

It seems that after the 1990s there were fewer cases of people entering the country in secret. However, there were some cases entering the country without passports around the year 2000. These included Person C, who became pregnant with her second child in Bangkok, and Person E, who stopped working abroad and returned to Laos.[21]

At any rate, based on the three cases presented in this section, it seems that if they had not been staying illegally, or if they had medical facilities nearby that they could use without issue regardless of their residency status, more of the women would have used contraceptive methods while away working in Thailand.

We can also note that the three women's cases all gained knowledge about contraceptive methods in the areas where they were working abroad. Person C and Person E learned about them from Laotian colleagues with whom they were working, while Person D learned about them from a doctor at a hospital. Furthermore, Person C knew by the time she gave birth to her first son in 1998, and Person D gained knowledge from a doctor in 1992. Person E knew about contraceptive methods by the time she got married in 1994. In other words, Person D and Person E learned about contraceptive methods in Thailand, where they worked, prior to the commencement of the Laotian government's family planning program in KH village in 1996.

[21] In Savannakhet Province, it was not unusual for this kind of situation starting in the 2000s. For example, in a survey conducted by Phouxay from 2004 to 2006 in two villages in Champhone District, Savannakhet Province, most of the people who had migrated to Thailand from both villages did not possess a passport or visa (Phouxay 2017: 381).

Meanwhile, it is unclear exactly when Person C learned about contraceptive methods. She first went to work at confectionery factory in Bangkok for two years, starting in around 1994. If we suppose that she learned this knowledge from friends around that time, it would place the event prior to 1996. Additionally, in Sect. 3, 10 of the women mentioned Bangkok as their source of information about contraceptive methods in the all-household survey. Furthermore, three of them, including Person A from Case 1, had started using contraceptive methods in the early 1990s.

In Sect. 4, we raised the question of how the women who started using contraceptive methods before the start of the government family planning program in 1996 came to start using these methods. Based on the previous and present sections, it can be posited that the opportunity to work abroad in Thailand was an important opportunity for them to start using contraception. Furthermore, we can position migrant work played an important role, particularly in the processes during the early stage of the use of contraceptive methods spreading among the women of KH village, in that it gave the women a chance to gain information about these methods.[22]

8 Conclusion

The purpose of this chapter was to reveal the processes and characteristics relating to the spread of the use of modern contraceptive methods in KH village and to consider the social context in which that spread took place. On these two points, we will close out this chapter by briefly summarizing what we have stated thus far, and adding some further considerations regarding the social context.

Firstly, regarding the processes by which contraceptive methods are disseminated, the following four characteristics can be identified.

(1) In KH village, contraceptive methods are exclusively used by women.
(2) Contraceptive methods came to be used widely in the late 1990s.
(3) Initially, contraceptive methods were used among women who had given birth and had multiple children, as means of control further births. However, they subsequently became used among women in younger age brackets as a method of planned birth control.
(4) The common methods used are the pill and injections.

Among these four characteristics, number (1) is in line with the tendency reported from other parts of Laos that women use modern contraceptive methods at a notably higher rate than men (Shimazawa 2014: 214; Chantakoumane et al. 2020: 77). Number (4) also corresponds to the characteristic found in NK village in northern Laos, discussed in Shirakawa and Moriki (2025). Meanwhile, in understanding the societal context that led to the spread of contraceptive methods in KH village, it

[22] The fact that migrant work plays a key role in the spread of family planning information was also brought up in studies in Guatemala and China (Lindstrom and Munoz-Franco 2005; Chen et al. 2010). However, these studies focused on migrant workers moving from rural to urban areas of the same country, rather than working abroad.

is impossible to ignore the Laotian government's family planning program, which began in 1996. Many women may have begun to use the pill or injections because of this. Furthermore, we can see that healthcare professionals and medical institutions played a significant role in the process of this spread. It resembles to the case of NK village where healthcare professionals also played a major role in the adoption of contraceptive methods among the women.

Nevertheless, a small number of women in KH village had started using contraceptive methods prior to the implementation of the program in 1996. They started to use it in the context of having gone abroad to Thailand for work. The use of contraceptive methods among the women of KH village gained momentum with the commencement of the government's family planning program, however, this trend originated with women traveling to Thailand for work. This process can be reflected a feature of KH village, that a large number of its residents have been engaged in migrant work. The process may also apply to other central Laotian villages located close to Thai border, such as the villages discussed by Phouxay (2017), where migrant work in Thailand is common.

The women of KH village who used contraceptive methods while working in Thailand, as stated in Sect. 6, did so to avoid issues they would face if they became pregnant. This leads to question of why the government's family planning program was accepted by the women of KH village. As mentioned in Sect. 3, the main reasons the women arrived at the use of contraceptive methods were the difficulty of childbirth, childrearing, and economic constraints they faced. Bearing these things in mind, the need to avoid childbirth, childrearing, and the economic burdens that would arise from becoming pregnant was widely shared by the women of KH village at the time the program was implemented. This may have been why the women accepted the program, as it could respond to their needs.[23] However, at present this remains a hypothesis. Future research will be required to verify the findings of this chapter, including examining the social and economic context that led to these women having the needs described above.

Acknowledgements This study was supported from JSPS Grants-in-Aid JP25257004. In writing this chapter, we received tremendous support from many people and related organizations. First, we would like to extend a heartfelt thanks to the people of KH village, who always tirelessly participated in the surveys that formed the basis of this chapter. When doing these surveys, Ms.

[23] When we conducted an informal interview in November 2015, female Nurse S at LN village health center, who runs activities for family planning in KH village (and whose mother resides there), stated, "In the past, having a lot of children was not appreciated, it was thought that it was better to have fewer children. However, at the time there was no pill, so there was nothing we could do about it. But now we have it, so everyone uses it." According to Nurse S, above all else, women did not want to have many children, and the need for using modern contraceptive methods had existed since the past. As Nurse S was born in 1986, supposing that "in the past" for her was her childhood, that would be the 1990s. In contrast to this, a woman from KH village in her late 70s (as of December 2014) remarked, "in the past, people were pleased to have a lot of children." Assuming that her "past" was when she was in her 20s, that would be the 1960s. If that is the case, and this is no more than conjecture, there were major changes in the value placed on having a large family by the 1990s, by which time it may have become common for women to want to avoid having many children.

Phatthana Phimmayoi and Ms. Sichan Pasanxi accompanied us and acted as intermediaries with the people of KH village. The support of the Savannakhet Provincial Health Department, Lao Tropical and Public Health Institute, and the National Agriculture and Forestry Research Institute was indispensable in conducting this survey. We would like to express our deep gratitude to everyone involved.

References

Aoyama A, Hara H, Kita E (2001) Health, development, and gender: well-being for all. Yūhikaku, Tokyo. (in Japanese)

Chantakoumane G, Maguet C, Essink D (2020) Married couples' dynamics, gender attitudes and contraception use in Savannakhet Province, Lao PDR. Glob Health Action 13:74–84. https://doi.org/10.1080/16549716.2020.1777713

Chen J, Liu H, Xie Z (2010) Effects of rural-urban return migration on women's family planning and reproductive health attitudes and behavior in rural China. Stud Fam Plann 41:31–44. https://doi.org/10.1111/j.1728-4465.2010.00222.x

Edmeades J (2008) The legacies of context: past and present influences on contraceptive choice in Nang Rong, Thailand. Demography 45:283–302. https://doi.org/10.1353/dem.0.0004

Entwisle B, Rindfuss RR, Guilkey DK, Chamratrithirong A, Curran SR, Sawangdee Y (1996) Community and contraceptive choice in rural Thailand: a case study of Nang Rong. Demography 33:1–11. https://doi.org/10.2307/2061709

Iinuma T (2013) Jendā byōdō to syakaisyugi soshiki: Kaihatsu kyōryokuka no Raosu josei dōmei no hensen wo tōshite (Gender equality and socialist organization: with respect to the transformation of the Lao Women's Union under the development cooperation regime). In Suzuki M, Takimura T (eds) Kokusai kaihatsu to kyōdō: NGO no yakuwari to jendā no shiten (International development and collaboration: the role of NGOs and gender perspective). Akashi shoten, Tokyo, pp 245–265. (in Japanese)

Kono S (2007) Jinkōgaku e no shōtai: Shōshi kōreika wa dokomade kaimei saretaka? (An invitation to demography: what is already known about declining birth rate and aging population?). Chūōkōronsha, Tokyo

Kiso K (2013) Providing support in child care for working mothers: living separately as a care practice in Northeast Thailand. Hikaku Kazokushi Kenkyū (J Comp Family Hist) 27:53–74 (in Japanese). https://doi.org/10.11442/jscfh.27.53

Kiso K (2016) Onna ga "totsugu" to iukoto: Tohoku tai nōson josei no dekasegi to Raifu kōsu no henyō (The meaning of woman gets married: labor migration and transformation of life course of female villagers of northeast Thailand). In Nakatani A, Udagawa T (eds) Shigoto no jinruigaku: Rōdō chūshinshugi no mukōe (Anthropology of work: beyond labor-centrism). Sekaishisōsha, Kyot, pp 92–110. (in Japanese)

Lao Statistics Bureau (2018) Lao social indicator survey II 2017. Vientiane

Lindstrom DP, Munoz-Franco E (2005) Migration and the diffusion of modern contraceptive knowledge and use in rural Guatemala. Stud Fam Plann 36:277–288. https://doi.org/10.1111/j.1728-4465.2005.00070.x

Matsuo M (2013) *Jendā to ripurodakushon no jinruigaku: Indo nōson shakai no funin wo ikiru joseitachi* (Anthropology of gender and reproduction: women living with infertility in rural village society of India). Shōwadō, Kyoto. (in Japanese)

Ministry of Health & Lao Statistics Bureau (2012) Lao social indicator survey 2011–12. Vientiane

Niwa T, Nakagawa S, Takahashi S, Nishimoto F (2021) Transnational labor migration from rural village in Laos to Bangkok, Thailand: focusing on relationship with the home village. Nenpō Tai Kenkyū (J Thai Stud) 21:17–30 (in Japanese)

Otsuka R (2015) *Hito wa kōshite fuetekita: 20 mannen no jinkō hensenshi* (The population growth of homo sapiens: 200,000 years of its demographic history). Shinchōsha, Tokyo. (in Japanese)
Phouxay K (2017) Patterns and consequences of undocumented migration from Lao PDR to Thailand. In: Bouté V, Pholsena V (eds) Changing lives in Laos: society, politics, and culture in a post-socialist state. NU S Press, Singapore, pp 350–373
Schenk-Sandbergen L (2012) The Lao matri-system, empowerment, and globalization. J Lao Stud 3(1):65–90
Shimazawa K (2007) Ripurodakushon: "Umukoto" wa tanjun dewa nainoka? (Reproduction: isn't it simple to give birth?). In: Ikeda M, Okuno K (eds) *Iryō jinruigaku no Ressun: Yamai wo meguru bunka wo saguru* (Lessons in medical anthropology: exploring cultures of illness). Gakuyōshobō, Tokyo, pp 148–171. (in Japanese)
Shimazawa K (2014) Raosu ni okeru "seishoku kontorōru" no yōsō: Josei no kenkō purojekuto toshite no dōnyū (Aspects of "reproductive control" in Lao PDR: its introduction as a health project for women). In: Kohama M, Matsuoka E (eds) Childbirth and family planning in Asia: the politics of women's reproductive bodies. Benseishuppan, Tokyo, pp 192–222. (in Japanese)
Shirakawa C, Moriki Y (2025) Aspects of modern contraceptive use among women in rural northern Laos. In: Yokoyama S, Niwa T, Jiang H (eds), Population dynamics and livelihood changes of small-scale societies in Laos. Springer, Singapore, pp. 207–222. https://doi.org/10.1007/978-981-96-2956-5_10
Sychareun V (2004) Meeting the contraceptive needs of unmarried young people: attitudes of formal and informal sector providers in Vientiane municipality, Lao PDR. Reprod Health Matters 12(23):155–165. https://doi.org/10.1016/S0968-8080(04)23117-2
Sychareun V, Phongsavan K, Hansana V, Phengsavanh A (2010) Policy maker and provider knowledge and attitudes regarding the provision of emergency contraceptive pills within Lao PDR. BMC Health Serv Res 10:212–221. https://doi.org/10.1186/1472-6963-10-212
Sychareun V, Hansana V, Phengsavanh A, Phongsavan K (2013) Awareness and attitudes towards emergency contraceptive pills among young people in the entertainment places, Vientiane City, Lao PDR. BMC Women's Health 13:14–23. https://doi.org/10.1186/1472-6874-13-14
Takahashi S, Yokoyama S, Nishimoto F, Niwa T (2023) Population growth and generational response to livelihood choices in a rain-fed farming village in central Laos. E-J GEO 18:264–278 (in Japanese). https://doi.org/10.4157/ejgeo.18.264
Tomita S, Badenoch N (2014) Tai bunkaken ni okeru tei jinkō zōkaritsu no kentō: 1971 nen kara 2006 nen ni okeru Raosu hokubu no isson no keiken kara (An investigation into low population growth rates in the Tay cultural area: the experience of a village in northern Laos between 1971 and 2006). In: Daniels C (ed) *Tōnan ajia tairiku sanchimin no rekishi to bunka* (The history and culture of upland peoples in mainland Southeast Asia). Gensōsha, Tokyo, pp 247–276. (in Japanese)
United Nations Population Fund (UNFPA) (2015) Family planning situation analysis: Lao people's democratic republic. UNFPA Lao PDR, Vientiane

Population Dynamics and Livelihoods in both Paddy and Swidden Farming Communities in the Basin

Population Dynamics and Paddy Holdings in Remote Rural Villages of Northern Laos

Satoshi Yokoyama, Shinichi Takahashi, Takahito Niwa, and Futoshi Nishimoto

Abstract This study investigated changes in population dynamics and paddy holdings over three generations in two villages established by the Lao people in a basin in northern Laos. In the study area, government policy has promoted migration from mountainous areas since the 1970s. However, there was limited land for new paddy reclamation in the basin, and therefore many residents who migrated from mountainous areas, mainly Khmu people, had limited opportunities to engage in wet-rice cultivation in the basin villages where they resided and ultimately continued the subsistence swidden agriculture they had practiced in the mountainous areas. The common method of acquiring new paddies in basin villages, where it is difficult to develop new paddies, is to purchase paddies from residents who have moved away from the villages. Many residents waited for such opportunities and worked in urban areas to make money to purchase paddies. Migrants from mountainous areas in these basin villages are faced with the choice of staying in the basin, where paddy fields are scarce, or migrating to urban areas. We found that the basin villages functioned as intermediate points for migration between mountainous and urban areas.

Keywords Population dynamics · Paddy holdings · Paddy trading · Urban–rural migration · Laos

This chapter was a revised version of Yokoyama et al. (2023).

S. Yokoyama (✉)
Graduate School of Environmental Studies, Nagoya University, Nagoya, Japan
e-mail: yokoyama.satoshi.i4@f.mail.nagoya-u.ac.jp

S. Takahashi
Kobe University, Kobe, Japan
e-mail: takahashi@econ.kobe-u.ac.jp

T. Niwa
College of Human and Social Sciences, Kanazawa University, Kanazawa, Japan
e-mail: niwa-2@staff.kanazawa-u.ac.jp

F. Nishimoto
Embassy of Japan in Laos, Vientiane, Laos
e-mail: nsmtfts@gmail.com

S. Yokoyama et al. (eds.), *Population Dynamics and Livelihood Changes of Small-Scale Societies in Laos*, International Perspectives in Geography 22,
https://doi.org/10.1007/978-981-96-2956-5_7

1 Introduction

This study focuses on residents who migrated from mountainous areas to basin areas of northern Laos; it clarifies population changes that occurred in the destinations of their migration and the relationship of these changes with the livelihoods of those who live there.

In northern Laos, with the exception of the indigenous Mon-Khmer-speaking Khmu people, the Tai-Kadai-speaking Tai people, the Hmong-Mien-speaking Hmong people, and Mien people have historically migrated repeatedly, living interspersed in southwestern China, northern Vietnam, and northern Thailand (LeBar and Suddard 1960: 35–43). As Halpern (1964: 9) states, "It can be truly said that almost none of the villages in Laos are really permanent, the few villages in the Vientiane area that have existed for several centuries being rare exceptions." Thus, the relocation of villages and the migration of populations are not rare events.

No major population movement were recorded during the French colonial era (Évrard and Goudineau 2004), but from the 1960s to the start of the 1970s, emergency evacuation village resettlements were observed because of aerial bombing during the Indochina War (Chazée 1999: 16–17; Goudineau 1997a: 11). Then, after the Lao People's Democratic Republic was established in 1975, ethnic minorities were forcibly relocated from mountainous and remote areas because of security concerns about the activities of armed groups and with the goal of reducing swidden agriculture (Chazée 1999: 17; Molland 2017). Goudineau (1997a: 13) notes that there were four development objectives parallel to these migration policies: "formulation of urban planning and national land planning," "protecting and profiting from natural resources," "integrating remote locations into the market economy," and "integrating minority ethnic groups into the national culture."

When the Lao People's Democratic Republic was established, the study area of northern Laos is scattered with a variety of ethnic groups, but since their settlements are small and they practice swidden agriculture and opium cultivation, the government views them as a deviant presence (Baird and Shoemaker 2007; Ireson and Ireson 1991). It is conceivable that the government's main development goals were to integrate these small ethnic minority settlements with Tai ethnic groups, such as the majority ethnic Lao people, and have them engage in paddy cultivation and use roads to access markets (Molland 2017).

Then, in the second half of the 1990s, the government promoted permanent settlement and advocated job-creation policies.[1] Furthermore, in 2004, the Politburo of the Central Party Committee of Lao PDR announced its view that villages should have no fewer than 500 people in the lowlands or 200 people in the highlands and ordered the consolidation of settlements (Baird and Shoemaker 2007). The purpose of this policy was to encourage resettlement and migration from the highlands to the lowlands. This resulted in the formation of many multiethnic settlements in the lowlands and in highland areas alongside main roads (Évrard and Baird 2017).

[1] In the Lao language, this is *chatsan asip khongthi*, which translates to "creating stable livelihoods."

However, migration policies that would put an end to highland swidden agriculture had been underway since before the start of the communist regime in 1975.*Unasylva*, published by FAO, includes the following article from 1957:

> "The majority of the peoples who inhabit the mountain regions of Laos practice shifting cultivation. This practice is particularly widespread in the north of the country, and the deforestation caused by forest clearing constitutes one of the principal problems of the Forest Service.
>
> This body, since its creation, has tried to abolish the practice. First, a school, intended to encourage permanent cultivation methods among the population, was created on the plateau of Xieng Khouang. Then, a policy of confining the areas where shifting cultivation was practiced was introduced.
>
> However, Laos has a fairly small population, and it is therefore still possible to find uncultivated fertile land. For this reason, the Forest Service from 1950 onward made attempts at resettling the hill tribes and establishing them on land suited to rice growing." (FAO 1957).

At the time that this article was published, Laos was still a kingdom that had not adopted communism. From the monarchical era onward, when ethnic Lao in the lowland practiced paddy cultivation, ethnic minorities in the highlands practicing swidden agriculture were criticized. Fox et al. (2009) note that in Laos, ethnic groups that practice swidden agriculture are categorized as ethnic minorities, migration policies are implemented to stop swidden agriculture, and the Department of Forestry's (Ministry of Agriculture and Forestry) power has expanded as forest protection has gathered momentum.

Previously, Goudineau (1997b) with support from the UNESCO and UNDP studied village resettlement from the highlands to the lowlands by conducting a large-scale survey of the population, livelihoods, infrastructure, education, health, and sanitation of 67 resettled villages in 22 districts in 6 provinces (i.e., Luang Namtha, Oudomxay, Xieng Khouang, Attapeu, Salavan, and Sekong) in Laos and wrote a report that revealed lifestyle changes before and after resettlement. In the resettled villages, it was reported that the mortality rate had risen because of residents' worsening health conditions, and there were many cases in which peoples' livelihoods were threatened because they had been compelled to change livelihoods. Furthermore, Lestrelin (2011) discussed state territorialization and the autonomy of ethnic groups based on the relationships between the Khmu people, which had migrated from the highlands, and the ethnic Lao in two villages that had been newly built alongside national road in the northern province of Luang Phabang. Baird and Shoemaker (2007) interviewed several development agencies in addition to reviewing existing literature. They discuss Laotian resettlement policies and have taken an extremely critical stance toward both government policies and the performance of development agencies. International aid organizations often participate in resettlement, and while some of them claim to support resettlement when it is voluntary, in the context of Laos, it is not easy to distinguish between voluntary and involuntary resettlement. Whether resettlement is voluntary or compulsory, highland Lao society has been subjected to dramatic deconstruction and reconstruction beyond its control within a short period of time. The issue of whether resettlement is voluntary or compulsory has also been debated in academic journals (High 2008; High et al. 2009).

Most existing studies have focused on the political aspects of resettlement and migration and have contributed significantly to raising awareness of these issues among members of the development community in Laos. However, for both voluntary and forced resettlement, there are no data-based studies on the lifestyles of migrants in resettlement areas or whether they settle permanently in those areas in the context of migration to the lowlands by those who practiced swidden agriculture in the highlands. There is also no research focusing on villages that receive these migrants. Observing only the areas of origin is insufficient when it comes to resettlement and migration; it is also necessary to observe the people and places at the destination.

Thus, this study focuses on a basin village in northern Laos that has received many migrants from mountainous areas and discusses the livelihoods of migrants in the basin village. The livelihood in the basin area of northern Laos is wet-rice cultivation in paddy fields with weir irrigation. Specifically, the study will clarify how ethnic minority migrants who have been engaged in swidden agriculture in the mountainous areas make a living in the basin villages by using data on population dynamics and the area of paddy fields owned by the migrants.

2 Study Area and Research Methods

2.1 Study Area

The study areas are NK and HB villages in Ngoi district, Luang Phabang province, in northern Laos, a landlocked country in mainland Southeast Asia (Fig. 1). The Ngoi district office is located in the Nong Khiaw area, but was formerly located in MN village on the Ou River basin, which was once a prosperous village. NK village is located about 5 km east of the former district office in MN village and about 7.5 km northeast of HB village. To reach NK village from MN village, one needs to cross an unbridged river in two places, but it is accessible with a four-wheel drive truck. However, to reach HB village, the water level of the Ngoi River, which runs north of NK village, is high and cannot be crossed by cars, so the only way to reach the village is by in a remodeled tractor or on foot. Furthermore, according to the most recent site survey in March 2023, no grid electricity has arrived in either village, each household generates its own electricity by installing hydroelectric generators. Cellphone coverage is also extremely weak, and these villages are located in the most underdeveloped area in northern Laos.

In addition to NK and HB villages, the study area includes the sites of NL, NT, and NM villages (Fig. 1). However, NT village was abandoned by 1970 and NL village by 1975 because all residents had migrated to Luang Phabang or Vientiane. The three villages of NK, HB, and NM continued to exist after Laos transitioned from a monarchy to a socialist country in 1975. However, about 15 households from NM village migrated to MN village in 2006 and 2–3 migrated to other parts of

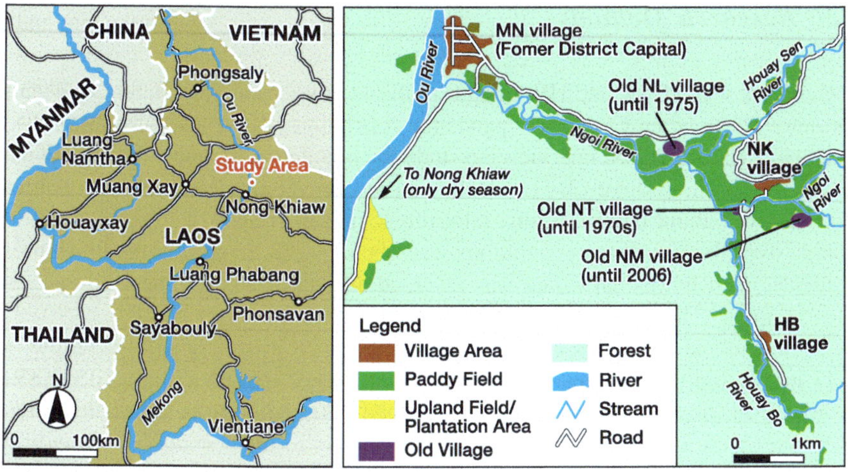

Fig. 1 Study area. *Source* Pleiades satellite data taken in 2017

Luang Phabang; thus, NM village was abandoned. Therefore, the only two remaining villages at present are NK and HB.

By the end of 2018, the population of NK village was 354, of which 63 people had temporarily left the village to work elsewhere. Out of 65 households, 43 were ethnic Lao, the largest ethnic group in Laos, and 22 were Khmu, the largest group ethnic group in northern Laos. According to NK village elders, the village has a long history and used to be called Na Onmong village.[2] It is unclear when it became NK village, but it can be traced back at least three generations and had been established about 100 years ago.

HB village had a population of 208, of which 27 people had temporarily left the village to work away. It had 42 households, of which 23 were ethnic Lao and 19 were Khmu. The history of HB village is complicated. In 1954, HB village saw increased Pathet Lao activity because of the influence of the communist revolution during the Laotian Civil War, and the village was abandoned after being burned to the ground by the opposing Royal Lao Army. The Royal Lao Army burned down the village with the aim of annihilating remote settlements that were difficult to monitor and forcing people to live in more densely populated settlements. Thus, the residents of HB village were forced to migrate to NT village. Thereafter, the Pathet Lao forces gained the upper hand, and when they came to control the entire study area, the Royal Lao Army began bombing the region. The residents of NT village temporarily lived hidden in caves to avoid the bombing. Some of the residents who had lived in NT village returned to HB village in the early 1970s, when the civil war was less intense; this was the generation of the current mayor's parents.

[2] *Na* means paddy, and *Onmong* was the name of a former queen; thus, the former name of the village was "Onmong's Paddy." The origin of the old village name lies in the legend that Queen Onmong died in this village about 460 years ago.

2.2 Research Methods

The authors visited NK and HB villages multiple times from September 2017 until February 2020, when COVID-19 pandemic restrictions prevented entry into Laos, and conducted structured interviews to study the population composition at the time of the all-household survey. By conducting unstructured interviews with each household member about deaths, births, relocations, and marriages, data on population change over the past three generations was obtained. To determine the kinship of residents, each individual's data was assigned an identifying number and entered into "Gramps",[3] an open-source genealogy application.

Simultaneously, high-resolution imagery data taken by the Pleiades[4] satellite in 2017 in NK and HB villages were used to input paddy field data into a GIS (ESRI ArcMap). In the study area, 102 people owned 200 paddy plots. Semi-structured interviews were conducted with paddy field holders to understand the changes in paddy holdings from the time of reclamation to the present. However, in the case of NK village, the information on paddy reclamation before three generations could not be identified because the village was established a long time ago. Therefore, the analysis of changes in paddy field holdings in this study is based on data from 1970s onwards, when the paddy field holders could be identified with certainty.

3 Population and Paddy Holding Changes

3.1 Population Changes

Roughly 50 years of population changes in NK and HB villages, from 1970 to 2018, are shown in Fig. 2.[5] Both villages show the tendency of continuous population growth.

A major event in terms of population change occurred in NK village in 1974, when 63 residents moved in from the mountainous and already abandoned Mok-Kou and Phon villages. Most of the in-migrants were Khmu people. In HB village,

---

[3] Gramps is an open source genealogy software program that stores parental, sibling, in-law, and other relationships in databases. Takahito Niwa and Futoshi Nishimoto created the study databases using Gramps. They can be downloaded here. https://gramps-project.org/blog/ (last viewed on January 16, 2023)

[4] Using 0.5 m panchromatic and 2 m multispectral 4-band ortho data.

[5] The population is estimated by tracing the genealogy of relatives back to the year 2020. Regarding the registered population, we attempted to obtain as much data as possible even for households that have already moved out or become extinct. However, there are some households that have been forgotten by the residents and could not be reconstructed. Regarding the real population, we excluded residents who had left the village to go to work or attend school outside the village while still holding a certificate of residence. As for the residents who have been away from the village for a long period of time and the time of their leaving is unclear, we estimated the population based on interviews with several residents.

Fig. 2 Population changes in the study area. *Source* Authors' interview survey

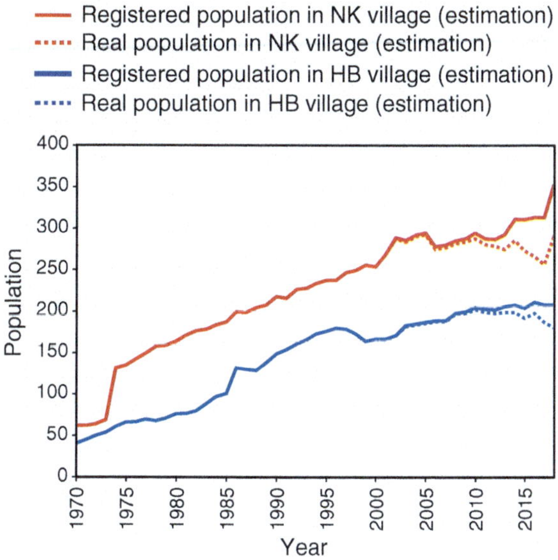

37 people moved in from Phonthong district, east of Ngoi district, in 1983 and 1986 (Fig. 3). In NK village, the influx of people from neighboring settlements has continued thereafter, particularly since 2000, with more than 140 in-migrants. A constant influx from neighboring settlements has also continued in HB village, a notable feature being that incomers to both villages have been mostly Khmu people.

Meanwhile, starting from the mid-1990s, the number of people leaving both villages has increased. In NK village, over 48 years, from 1970 to 2018, 275 people have moved in, while 143 people have moved out (Fig. 3a). In HB village, 165 people have moved in and 139 have moved out (Fig. 3b). Additionally, there are residents

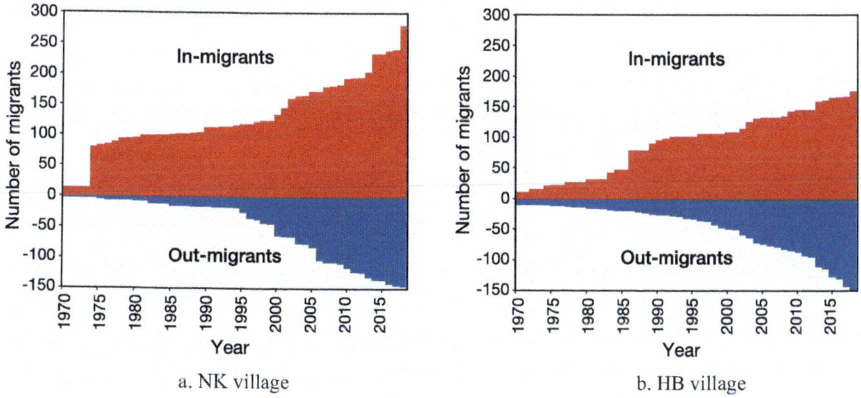

Fig. 3 In-migrants and out-migrant in the study area **a** NK village **b** HB village. *Source* Authors' interview survey

with certificates of residence in the villages, but many have left temporarily; the difference between the dotted and solid lines in Fig. 2 shows the number of villagers who have left temporarily. Most of those who have left temporarily are pursuing migrant work in urban areas. Ten or more people per year began to leave temporarily starting in 2013 in NK village and in 2015 in HB village, and these numbers have continued to grow in both villages.

3.2 *Paddy Field Distribution and Holding*

In basin villages like those in this study area, there continues to be an increase in residents moving in from mountainous areas. However, are the migration destinations able to provide incoming populations with livelihoods that enable them to secure enough food?

Agriculture is the main livelihood in northern Laos, where commerce and industry are undeveloped. Subsistence rice farming is the foundation, with mountainous areas engaging in swidden upland rice cultivation and basin areas, like the study area, building river weirs, drawing water to paddy fields, and engaging in wet-rice cultivation (Fig. 4). Non-timber forest products (NTFPs) are also gathered (Openshaw and Trethewie 2006; Thanichanon et al. 2013; Yokoyama 2004), and livestock like cattle, pigs, and poultry are raised, providing an important source of cash income for subsistence farmers (Phouyyavong et al. 2020; Yokoyama 2004).

Most Laotian farmers prefer wet-rice cultivation in paddy fields, which gives stable yields, instead of swidden upland rice farming, which is influenced by the weather. With swidden agriculture, fluctuations in precipitation during the rice crop's development period affects the yield, and the cutting and burning season must be ascertained through experiential knowledge. For example, Ingxay et al. (2015a, 2015b) discuss that in northern Laos, in 2011, when the rainy season started earlier than usual, many swidden farmers were left destitute because they could not start fires in many areas and were unable to make the land arable with swidden agriculture. The practice of

a. Weir irrigation b. Paddy fields

Fig. 4 Weir irrigation and paddy fields in NK village, 2017

wet-rice cultivation in paddy fields is stabler than upland rice farming in swidden field in relation to such aspects of climate change. Therefore, regardless of whether their migration from mountainous to basin areas was compulsory or voluntary, migrants could expect to get paddy fields in their destinations, which was a motivational factor. However, the basin areas in which the study areas of NK and HB villages are located are unable to provide enough paddy lots.

Figure 5 shows the places of residence of paddy holders in the study area. Many paddy fields have been reclaimed in the surroundings of NK village, and their area extends to 84.27 ha, but their owners are not only the residents of NK village. MN village, which was positioned long ago as the central settlement in this area, reclaimed paddy fields along the Ngoi River (see Fig. 1), and its eastern end extends into what is now the west side of NK village territory. The area of MN village paddy fields, shown in Fig. 5, is a total of 39.98 ha. Of these, 17.64 ha of paddy fields belong to residents of the former NM village, who continued to own them even after migrating to MN village. The former NM villagers did not relinquish these paddy fields, and they commute from MN village to work on their fields. Therefore, NK villagers hold no more than roughly half, or 42.25 ha, of the paddy fields within the territory of their own village. Meanwhile, 25.2 ha of paddy fields have been reclaimed in the territory of HB village, and their ownership is limited to the residents of the village.

3.3 *Paddy Reclamation*

Table 1 shows the paddy area cultivated in the territories of NK and HB villages, as well as the paddy area ratio in 1969 and 2018.In the territories of both villages, over 80% of the paddy area currently in use had already been reclaimed as of 1969. That is, before the arrival of more migrants from mountainous areas in 1969, most of the paddy reclamation had been completed. These paddy fields were reclaimed two or more generations ago and have been inherited by members of the families living in the study area. As mentioned above, NM village residents who migrated to MN village in 2006 kept their paddy fields even after migrating and continue to cultivate them.

According to the 4th Lao Expenditure and Consumption Survey, conducted by the Lao Statistics Bureau in 2007–2008, residents of rural areas have an average intake of 545 g of rice, that is, approximately 199 kg of rice is consumed per year (Committee for Planning and Investment, Lao Statistics Bureau 2012). The hulled rice yield of paddy in the study area was roughly 2.65 t/ha (roughly 3.78 t/ha for unhulled rice) in 2018 according to the results of the all-household survey conducted in HB village in September 2019.[6] Therefore, to secure the average rice consumption

[6] We interviewed all households on their 2018 rice harvest, and the responses provided the number of bags harvested. A bag of unhulled rice weighs about 60 kg, while polished rice is roughly 70% of that, at about 40 kg. Yields are averaged values obtained by dividing the yields of households owning paddy fields by the paddy field area as measured with GIS.

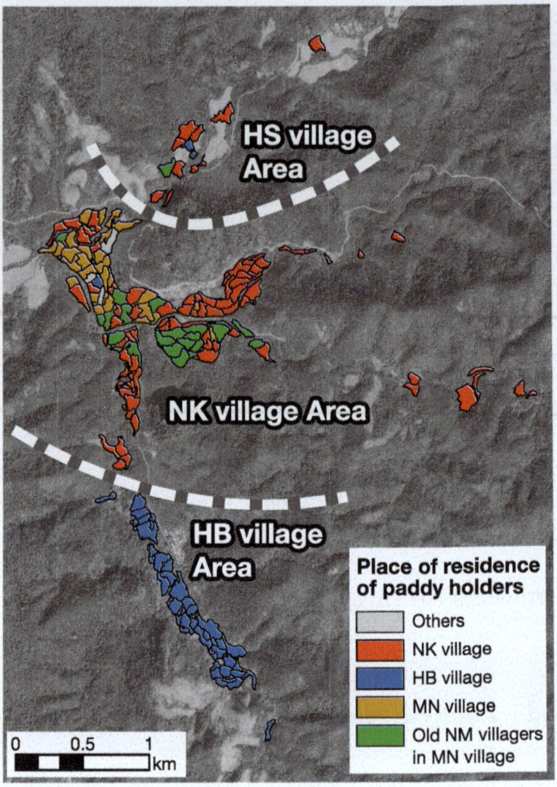

Fig. 5 Place of residence of paddy holders in the study area. *Source* Pleiades satellite data taken in 2017 and authors' interview survey

Table 1 Area of paddies in 1969 and 2018

Village	Area in 1969 (ha)	Area in 2018 (ha)	Area ratio (1969/2018)
NK village	68.63	84.27	0.81
HB village	21.10	25.15	0.84

Source Authors' interview survey

volume of 199 kg/year, which is consumed in rural areas, a paddy holding area of approximately 751 m^2 per-person is required.[7]

Figure 6 shows the reclaimed paddy areas from 1970 onward and the paddy area per-person as the targets of analysis. In NK village, there were 27.63 ha of paddy fields owned by a population of 62 people in 1970; that is, each person owned an average of 4,457 m^2 of paddy fields (Fig. 6a). A large number of residents moved into NK village from mountain villages in 1974, with the population swelling to roughly doubled, and the per-person paddy holding area dropped to 2,160 m^2. The migrants looked for suitable paddy lots in the village and reclaimed 9.40 ha of land on the

[7] Rice intake does not take age into account and is the value divided by the 2018 population.

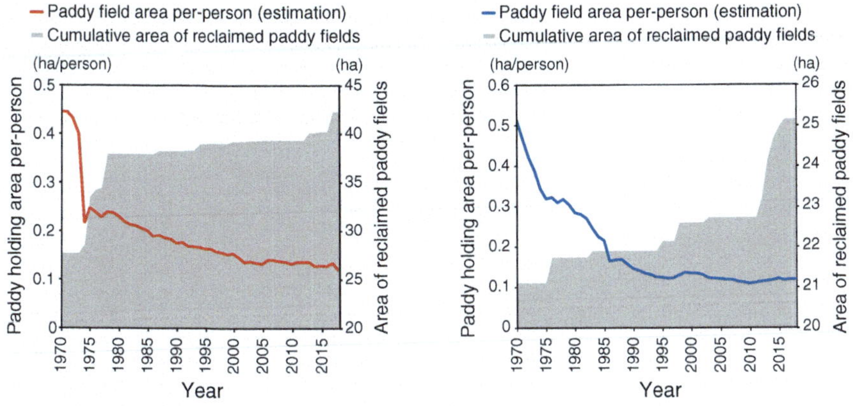

a. NK village

b. HB village

Fig. 6 Paddy holding area per-person and the cumulative area of reclaimed paddy fields. *Source* Authors' interview survey

eastern side and the northern extremities of HK village over a four-year period from 1975 to 1978. After that, during the following 38 years from 1979 to 2018, the only paddy fields that were reclaimed were 4.34 ha in the study area (Fig. 7). Even so, the per-person paddy holding area in 2018 was 1,197 m^2; the paddy area surpassed the average rice consumption volume in Laotian rural areas.

In HB village, 21.10 ha of paddy fields were owned for a population of 41 people as of 1970; that is, each person owned an average of 5,147 m^2 of paddy fields (Fig. 6b). However, as the village accepted migrants from the mountainous areas and the population grew, the per-person paddy area continued to fall. In HB village, limited lots were suitable for paddy fields, and the flanks of existing paddy fields have only been minimally developed (Fig. 7). Only 4.05 ha of paddy field has been reclaimed since 1970. However, the per-person paddy holding area in 2018 was 1,209 m^2, and similar to NK village, the paddy holding area exceeds the average rice consumption in rural Laos.

From the 1970s onward, almost no plots could be reclaimed as paddy fields in the study area, but even as the population has grown, village residents have been able to harvest more than roughly 199 kg of rice per year, which is the average rice intake for rural Laos. A reason that made it possible for basin area settlements to receive large numbers of migrants from mountain areas may have been the presence of highly productive paddy fields that have been worked for a long time.

Consequently, while rice subsistence was achieved in the study area, paddy fields were not distributed equally to migrants; longstanding residents continue to own them; and there is no land left for newly migrating residents to reclaim as paddy fields. Therefore, large discrepancies have arisen between households in the village in regard to paddy holding area. Thus, when we conducted a statistical test on ethnic differences in paddy holding in the villages. The nonparametric Wilcoxon rank-sum test was used, because paddy ownership in the study area has a large standard

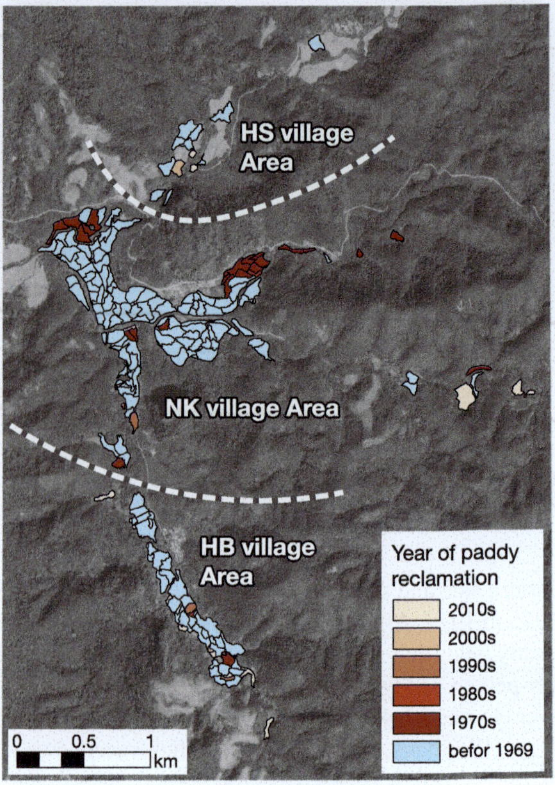

Fig. 7 Distribution of paddies by year of reclamation in the study area. *Source* Pleiades satellite data taken in 2017 and authors' interview survey

deviation and cannot be regarded as a normal distribution. The result showed that while there was no significant difference between ethnic Lao and Khmu people in NK village, a significant disparity had arisen in HB village (Table 2).

Table 2 Paddy holding in NK and HB villagers in 2018

	NK village $n = 65$HHs		HB village $n = 42$HHs	
	Ethnic Lao (ha) $n = 43$HHs	Khmu (ha) $n = 22$HHs	Ethnic Lao (ha) $n = 23$HHs	Khmu (ha) $n = 19$HHs
Average paddy area	0.68	0.78	0.93	0.25
Standard deviation	0.75	0.50	0.76	0.37
p value$^{(1)}$	0.21		0.0015**	

Source Authors' interview survey
Note (1) ** indicate significant differences between two groups by Wilcoxon rank-sum test ($p < 0.01$)

In NK village, the Khmu households that migrated there in the 1970s had reclaimed a large area of land as paddy fields. They also bought up the paddy fields of households moving to other villages, increasing their ownership area; resultantly, as of 2018, the Khmu have a greater average paddy holding area than the ethnic Lao. However, in HB village, Khmu acquisition of paddy fields has not progressed.

Furthermore, based on the data given in Table 2, there is a large standard deviation in the paddy holding area of ethnic Lao households. While small in number, there have been ethnic Lao people among the migrants coming in from mountainous areas, and like the Khmu people, the migrant ethnic Lao people were unable to acquire paddy fields. This explains the large standard deviation; the ethnic Lao who came to NK or HB village long ago own large tracts of paddy fields, whereas the newest migrant ethnic Lao households have been unable to acquire paddy fields.

4 Methods of Acquiring Paddy Fields

4.1 Methods of Acquiring Paddy Fields in the Study Area

In the study area, which does not have any major industries, holding paddy fields and cultivating rice to obtain one's staple food is considered the most important activity. Ethnic Lao households that have been living in the area for a long time have been working the paddy fields over the course of several generations. These fields have been handed down among relatives. Meanwhile, the newly immigrated Khmu people practice wet-rice cultivation by reclaiming paddy fields themselves, acquiring paddy fields through trade, or renting paddy fields as tenant farmers.

Table 3 shows paddy acquisition methods in the study area as of the end of 2018. "Inherit" indicates paddy fields acquired through inheritance after having been reclaimed by an ancestor (i.e., biological or non-biological parents, etc.). All the paddy fields were reclaimed by someone, but those that were reclaimed personally during or after 1970 are labeled "Reclaim." "Inherit + Purchase" refers to paddy fields passing hands after being purchased by a family predecessor and includes inherited paddy plots that was traded among siblings.

The ethnic Lao households that founded the NK and HB villages own 81% of the paddy fields in the study area, and newly arrived Khmu households only own 19% of the paddy fields. The most common method for acquiring paddy fields is inheritance, accounting for 39%. As of 2018, since over 50 years have passed, the immigrant Khmu people handed down 4.16 ha of the paddy fields they reclaimed in the 1970s to the next generation. The next most common method for acquiring paddy fields is purchasing; 37.47 ha of paddy field was acquired this way. These purchases will be analyzed in detail later.

Comparing the paddy acquisition methods of the ethnic Lao people, who have resided in the area for a long time, with those of the Khmu people, who immigrated

Table 3 Paddy acquisition patterns in the study area, 2018

Paddy acquisition patterns	Area (ha) [(1)]		
	Ethnic Lao	Khmu	Total (Ratio %)
Inherit	41.57	4.16	45.73 (38.84)
Purchase	26.92	10.55	37.48 (31.83)
Reclaim	7.73	4.86	12.59 (10.70)
Lease	9.56	1.56	11.12 (9.45)
Purchase + Inherit [(1)]	9.02	1.80	10.82 (9.20)
Total (Ratio%)	95.80 (80.52)	21.93 (19.48)	117.73 (100)

Source Authors' interview survey
Note (1) The data include cases in which inherited paddies were traded, as well as cases in which inherited paddies were traded among siblings

from mountainous areas, the ethnic Lao people acquired paddy fields through inheritance and purchase, whereas the Khmu people most often purchased, then reclaimed, and finally inherited land. However, all the paddy fields that the current generation of Khmu people inherited were reclaimed by the previous generation. Ultimately, the only ways for migrants to acquire paddy fields are to reclaim them themselves by seeking suitable paddy sites or to purchase them from someone who is selling them. In Laos, a country that has implemented policies to promote migration from mountainous areas to basin areas, it is difficult for migrants to practice wet-rice cultivation in their new villages. Below, we will analyze paddy acquisition by the Khmu people who migrated from mountainous regions since 1970.

4.2 *Paddy Acquisitions by Khmu Migrants*

We analyzed the distribution of paddy fields owned by Khmu people who had moved into the study area and their methods for acquiring those paddy fields (i.e., reclamation, tenant farming, or purchase), as of 2018 (Fig. 8).

Nearly all the paddy field reclamation by the immigrant Khmu people happened in the 1970s, with no fields being reclaimed in the period from the 1980s to the 2000s.That is, those who were able to reclaim lands suitable for paddy cultivations were households that moved to the area in the early part of the 1970s. However, starting in 2010, Khmu immigrants proactively reclaimed the limited land that remained. These plots flanked existing paddy fields or were far from the settlement and near the source of the river. This reclaimed area was approximately 7.18 ha

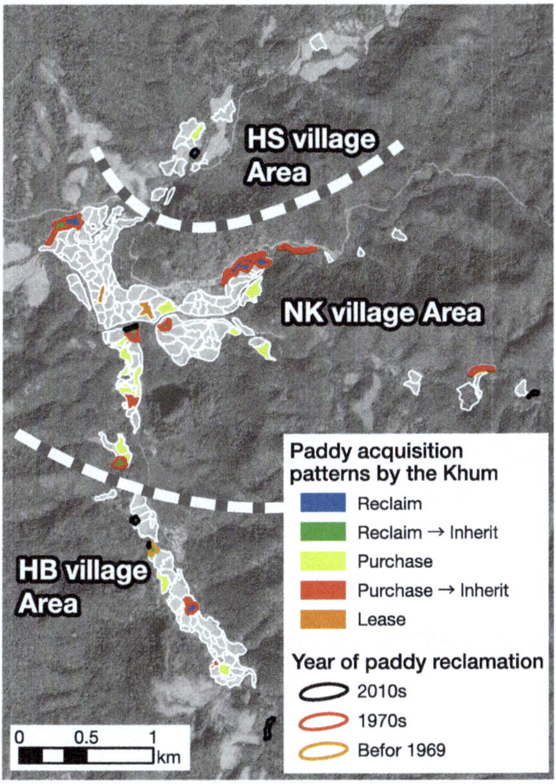

Fig. 8 Paddy holdings and their acquisition patterns by the Khmu in 2018. *Source* Pleiades satellite data taken in 2017 and authors' interview survey

(17 plots), held by 12 households in NK village territory; approximately 1.63 ha (6 plots) held by 4 households in HB village territory; and 0.21 ha (1 plot) held by 1 household in HS village territory. Note that there was only one plot in HB village territory that was reclaimed before 1969 and then inherited, but this was inherited by a Khmu man who had married an ethnic Lao woman from HB village; he inherited the paddy field from his father-in-law.

The only examples of Khmu tenant farming were two households renting about 1.56 ha (two plots) in NK village territory from a landowner in MN village (Table 3, Fig. 8). One plot was a 0.57 ha of the paddy field owned by Person T, a teacher who had moved to MN village from Vientiane Province. Person T purchased the paddy field from another resident of MN village (period unknown) and worked the field on their own, but since 2016, they began to lend it out to Khmu Person L of NK village. Person L was born in 1989 and moved to NK village from a mountain village in 2000. Person L pays the landowner, Person T, half of the rice harvested as tenant farming rent.

One more plot was a 1.00 ha of the paddy field owned by Person C who had lived in MN village for a long time. Person C began renting out the paddy fields passed down by their parents to Khmu Person X of NK village since 2004. Born in 1956,

Person X moved to NK village from a mountain village in 1974. Afterward, he met his wife, another Khmu person who had immigrated to the area like him, and they had six children. At present, four of the children have moved away from NK village to work in Vientiane or Thailand. As a tenant farmer, Person X receives 1.2 t of unhulled rice, and the rest is the share of the landowner, Person C. Considering that the unhulled rice yield of this region is roughly 3.8 t/ha, the rice distribution ratio would be roughly one third for the tenant farmer and two thirds for the landowner. Landowner C died in 2016, and his wife, Person B, is the current owner, but since Person C's death, she lives in Vientiane.

4.3 Acquiring Paddy Fields Through Purchase

Regarding the purchase of paddy fields by Khmu households, 11 households in NK village purchased 9.36 ha and 5 households in HB village purchased 2.99 ha (Table 4). There is limited land suitable for paddy cultivation in the study area, and although there are many households that would like to purchase paddy fields, few households are releasing the paddy fields they have inherited from past generations, so there are very few opportunities to purchase paddy fields. Not only the Khmu but also ethnic Lao households that have been living in the area for a long time want to purchase paddy fields. Recently married young couples, who are starting new households, have a particularly hard time gaining ownership of paddy fields, which forces them to lead difficult lives. Therefore, we need to shift our attention to include not only Khmu migrants but also to the buying and selling of paddy fields by all residents in the study area, where it is difficult to reclaim paddy fields.

From 1970 to 2018, residents of the study areas who had acquired paddy fields through purchase comprise 16 of the 27 ethnic Lao households and 11 of the 18

Table 4 Households that acquired paddies thorough purchase in NK and HB villages, 1970–2018

	NK village (*n*=65HHs) Area (ha) [No. of HHs]		HB village (*n*=42HHs) Area (ha) [No. of HHs]		
	ethnic Lao (*n*=43HHs)	Khmu (*n*=22HHs)	ethnic Lao (*n*=23HHs)	Khmu (*n*=19HHs)	Total (*n*=107HHs)
A) Paddies owned by villagers 1)	29.87 [27]	16.53 [18]	21.97 [17]	4.20 [8]	73.22 [70]
B) Paddy trading2)	13.24 [16]	9.36 [11]	8.53 [11]	2.99 [5]	34.12 [43]
a) with residents who moved to other areas	10.68 [14]	5.47 [6]	6.75 [8]	2.04 [3]	24.94 [31]
b) within villagers in the study area	2.57 [5]	3.89 [6]	1.78 [4]	0.95 [2]	9.20 [17]

Source: Authors' interview survey

Note: 1) Excluding paddies owned by villagers outside of NK and HB villages.

2) In some cases, the sum of a) and b) does not equal B, because one household trades paddy fields with several residents. The data include cases in which inherited paddies were traded, as well as cases in which inherited paddies were traded among siblings.

Khmu households of NK village and 11 of the 17 ethnic Lao 17 households and 5 of the 8 Khmu households of HB village (Table 4).

As shown in Table 3, 32% of the paddy fields worked in the study area were acquired through purchase. However, as shown in Table 4, if we limit it to residents of the NK and HB villages, the proportion grows considerably to 34.12 ha of the paddy area having been purchased from among the total 73.22 ha area owned or 47% of the paddy area having been acquired through purchase.

In both the NK and HB villages, the Khmu people have acquired less paddy area through purchase than the ethnic Lao people. In HB village, especially, because it is difficult to newly reclaim paddy fields and Khmu households were unable to acquire paddy fields through purchase, a disparity has arisen in the paddy area owned by ethnic Lao and Khmu people (Table 2).

Furthermore, of the 43 households that acquired paddy fields through purchase, 31 bought paddy fields from households who were letting them go after moving from the study area to other regions. This accounts for 73% (24.94 ha) of the purchased paddy area. Most of the households that have let go of valuable rice paddy fields have left the village, with a low rate of 27% (9.20 ha) of the paddy fields buying and selling occurring between village residents. The households from NK village that sold their paddy fields and moved to other areas went to Luang Phabang, Nong Khiaw, and Nam Bak districts in Luang Phabang province, while their counterparts from HB village moved to Vientiane (Fig. 1).

In the basin villages that welcomed many migrants from mountain areas, there is limited land suitable for paddy fields, used to obtain the staple food of rice, which is essential for survival; further, paddy reclamation is difficult. Therefore, purchasing the few paddy fields that belonged to households that moved out of the study area is a way to acquire remaining paddy fields.

5 Wet-Rice Cultivation and the Livelihood Structure of HB Village

This section focuses on HB village, where the Khmu people own paddy fields at a lower rate than ethnic Lao, and analyze the results of a household survey we conducted on how Khmu people maintain their livelihoods in circumstances in which they cannot procure paddy fields. We conducted the household survey in September 2019, obtaining 2018 livelihood data from 21 ethnic Lao households and 17 Khmu households, for a total of 38 of the 42 households of HB village.[8]

[8] It has 42 households, of which 23 are ethnic Lao and 19 are Khmu. We were unable to get data from one ethnic Lao household and one Khmu household that were not present during the survey period. Furthermore, we excluded two households from the analysis, one an older single-person ethnic Lao household (Person XY) and the other a Khmu household that was working in Vientiane in 2018, which did not engage in rice cultivation. Ethnic Lao Person XY lives on remittances received from children and relatives working in other areas. Therefore, we analyzed 21 of the 23 ethnic Lao households and 17 of the 19 Khmu households. Note that three of the ethnic Lao households do not

5.1 Ethnic Groups and Rice Cultivation

The classification of ethnic groups in the study area was determined by the father's ethnic origins because children inherit the ethnic group of their father, who is the head of the household. Therefore, the child of an ethnic Lao husband and a Khmu wife would be classified as ethnic Lao and the child of a Khmu husband and an ethnic Lao wife as Khmu.

HB village was originally an ethnic Lao village, and Khmu people began to immigrate there in the early 1970s, during the Laotian Civil War. Of the 19 Khmu households, five household heads were born in HB village; however, their fathers were Khmu people who had migrated there. The first household arrived in 1972, then 11 households came in the 1980s, 1 in the 1990s, and 5 more in the early 2000s. Since the current 19 households include 1 new household that just became independent, there were 18 households at the time of the migrations. Of these, only eight households own paddy fields, all of which acquired them through purchase (Table 4).

Meanwhile, 9 of the 23 ethnic Lao households migrated to the village in 1980 or later. Three households arrived in the 1980s, two in the 2000s, and four in the 2010s. The remaining 14 households were the initial settlers, who built the village in the early 1970s, and their children. With these 14 households at the core, the addition of new households from outside made HB village what it is today. Only three ethnic Lao households do not own paddy fields.

Figure 9 shows household rice harvest by ethnicity and rice production mode as of 2018. While the average rice harvest for the 21 ethnic Lao households was 85 bags, the average rice harvest for the 17 Khmu households was less than half of this, at only 37 bags. The ethnic Lao and Khmu rice harvest differs at the 1% level of significance in the Wilcoxon rank sum test (Fig. 9a).

In HB village, rice cultivation is practiced in both paddy field and swidden field. We observed three rice production modes being practiced: wet-rice cultivation in paddy field only, upland rice cultivation in swidden field only, and a mixture of both types. Of the 21 ethnic Lao households, the only one that was practicing swidden agriculture alone was a household that moved to the village in 2008 (Fig. 9b). There were 14 households practicing paddy field cultivation only, while 6 others had paddy fields but did not harvest enough rice from their paddy fields alone because the area of paddy fields they owned was too small; thus, they also practiced swidden agriculture. Some of the ethnic Lao households practicing both types of cultivation had greater crop yields than households that practiced only in paddy field. However, since there was only one ethnic Lao household that was engaged in upland rice cultivation in swidden field only in three forms, a Steel–Dwass multiple comparison test was no significant difference among the three cultivation types and the rice crop yield.

As for the Khmu, the average rice crop harvest engaged in paddy field only was 35 bags, which is extremely low compared to the average of 91 bags harvested by ethnic

own paddy fields but use their parents' paddy fields to cultivate rice. Therefore, the actual number of paddy holders is 17 households, but 20 households practice paddy farming.

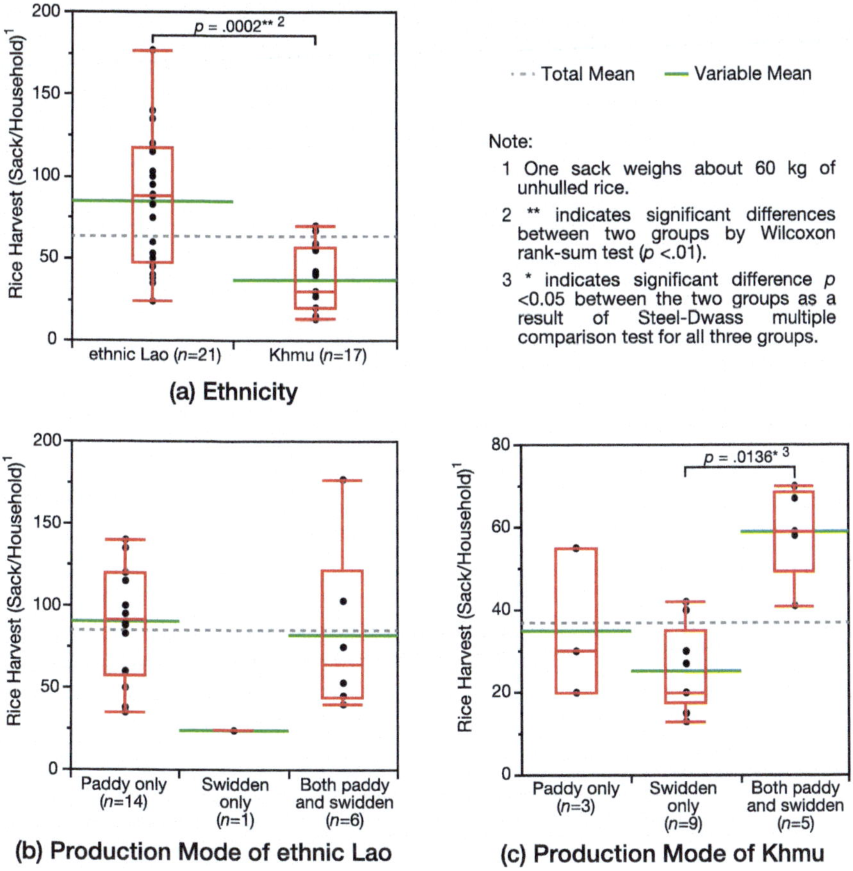

Fig. 9 Household rice harvest by ethnicity and rice production mode of NK and HB villages in 2018. *Source* Authors' interview survey

Lao households (Fig. 9c). Comparing the rice crop yields of ethnic Lao and Khmu households practicing both types of cultivation, the largest rice crop harvest of the Khmu people was 70 bags, which is less than the Lao average of 82 bags. Households practicing rice cultivation only in swidden fields harvested very little rice, with an average of 25 bags. Steel–Dwass multiple comparison tests were conducted among the three Khmu groups, and the results showed that there was a difference in rice crop harvest between the two groups, swidden field only and a mixture of both types, at the 5% significance level. This suggests that households engaged in rice cultivation in paddy fields only and swidden fields only do not differ in the rice crop harvest, but they secure the necessary amount of rice by combining both types of rice cultivation. In particular, there is concern that households that do not hold paddy fields and only cultivate rice in swidden fields may be lacking in rice needed.

Table 5 Village of origin and paddy holdings of married couples in HB village, 2018

	Birthplace of the couple		
	Both are HB village	Either of couples is HB village	Both are other village
Ethnic Lao (Household that hold paddies)	4 (4)	12 (12)	5 (4)
Khmu (Household that hold paddies)	0 (0)	9 (4)	8 (4)

Source Authors' interview survey

5.2 *Marriage Status and Paddy Field Holding*

In HB village, there are many residents migrated from other villages or migrated for marriage. If both spouses, or either husband or wife, are born in the HB village, they are expected to inherit paddy fields from their parents. Table 5 shows that the relationship between the origin of the married couples and the paddy field holdings. Only one ethnic Lao household did not hold paddy fields, and that was the only household where both spouses were from other villages. Among the Khmu, none of the married couples were from HB village, four of the nine households where one of the spouses was from an HB village, and four of the eight households where both were from other villages owned paddy fields.

Among the ethnic Lao and the Khmu households, eight households had paddy fields, and three households acquired their paddy fields by purchase, two households reclaimed the paddy fields themselves, and three households inherited them. The three households from other villages who had inherited paddy fields showed a variety of acquisition methods. One Khmu household migrated to HB village with their parents and inherited paddy fields that their parents purchased, one ethnic Lao household inherited paddy fields that their parents reclaimed, and one Lao household migrated from another village to HB village as an adopted child and inherited her paddy fields from foster father.

Of the eight marriages between the Khmu and the ethnic Lao in HB village, only one Khmu household inherited paddy fields reclaimed by Lao parents, as described earlier in Sect. 4.2. Accordingly, it is difficult to conclude for the Khmu households that intermarriage has changed the condition of paddy field holdings.

5.3 *Cash Income Activities*

The Khmu households of HB village, who hold limited paddy areas, need to engage in activities other than rice cultivation to earn cash for purchasing rice. The cash income activities in 2018 are shown in Table 6. The average household income was

Table 6 Income-generating activities of HB households in 2018

Activities	Average income (Kip)		p value[3] between two groups
	Ethnic Lao (n = 21HHs)	Khmu (n = 17HHs)	
(a) Rice	1,047,619	179,411	0.0159*
(b) Rice liquor	266,666	35,294	0.2304
(c) Field crops	102,380	86,470	0.7622
(d) NTFPs	924,738	2,321,764	0.0284*
(e) Intermediating	314,761	58,823	0.6378
(f) Handicraft	900,000	168,823	0.0064**
(g) Livestock	7,353,809	1,002,647	<0.0001**
(h) Others[1]	828,571	820,588	0.8201
Total[2]	11,471,881 (US$ 1,350)	4,638,529 (US$ 545)	0.0042**

Source Authors' interview survey

Note (1) Others include government allowances, remittances, tourism income, and sales of game from hunting

(2) The exchange rate for the Lao Kip in September 2019 was 8,500kip for US$1.

(3) In the Wilcoxon rank-sum test between the two groups, ** indicates a significant difference $p<0.01$ and * indicates a significant difference $p<0.05$.

11.47 million kip (US$1,350) for the ethnic Lao and 4.64 million kip (US$545) for the Khmu, with the ethnic Lao households earning 2.47 times more cash income than the Khmu households. The Wilcoxon rank-sum test showed a significant difference in average household income between the two ethnic groups at the 1% level of significance. According to the 6th Lao Expenditure and Consumption Survey 2018/19, conducted by the Lao Statistics Bureau in 2018–2019, the agricultural income (i.e., rice, field crops and fruits, meat, fish and shellfish, timber and NTFPs, and other) of rural households with difficult access to roads was 16,489,000 kip (Lao Statistics Bureau 2020: 45). Compared with the average agricultural income of rural Laotian households facing difficulties in accessing roads, the average income in HB village reflects extremely low levels of cash income, with ethnic Lao people reaching about 70% and Khmu people about 30% of that level.

5.3.1 Rice Sales

Eleven of the 21 ethnic Lao households sold rice, all of which were households that only practiced wet-rice cultivation in paddy fields. Meanwhile, only three of the 17 Khmu households sold rice, of which one practiced only wet-rice cultivation and the other two practiced both paddy and swidden rice cultivation. Households practicing only swidden agriculture did not sell any rice. Thus, only households in possession of paddy fields could sell rice, and ethnic Lao having a higher percentage of paddy holding got greater income from rice sales, with a significant difference at a level of

5%. However, the majority of the rice was consumed for subsistence, and even when there was a rice surplus, it did not contribute in a large way to cash income.

5.3.2 Liquor Sales

Five households (four ethnic Lao and one Khmu) earned income from the sale of home-distilled rice liquor. All of these were households that practiced paddy farming only and were also selling rice. The profit from distilling rice into liquor is about twice that of selling it as is. However, not everyone can make liquor. This is because a rice fermentation starter is needed to make liquor. The techniques for making rice fermentation starter are generally handed down from generation to generation within households, and not everyone gets to learn these techniques. One of the ethnic Lao households selling liquor was earning more money from selling rice fermentation starter than from selling distilled liquor. Only five households were selling liquor, and there were no significant differences seen between ethnic groups.

5.3.3 Field Crop Sales

Northern Laos is known for banana plantations (Hua et al. 2022; Friis and Nielsen 2016); rubber afforestation (Friis et al. 2016; Junquera et al. 2020; Lu 2017; Manivong and Cramb 2008; Thongmanivong et al. 2009); hybrid maize cultivation for animal feed (Cole 2022; Kallio et al. 2019); and contract cultivation of various crops intended for China (Fullbrook 2007; Hua et al. 2022), but none of these field crops or rubber afforestation was found in the study area. The reason contract cultivation was not seen in the study area was that the transport infrastructure for shipping crops was not in place. In HB village, there were no households cultivating field crops by contract, and their income from such crops was extremely low.

The crops sold by the residents of HB village were green onions (one household), carrots (one), cabbage (one), coconuts (two), maize (two), bananas (one), and tobacco (eight), all of which generated little income and did not differ significantly between the two ethnic groups.

5.3.4 Gathering NTFPs

The gathering NTFPs is a lively economic activity practiced regardless of ethnicity by 16 out of 21 ethnic Lao households and 16 out of 17 Khmu households. The species gathered include galangal seeds (*Alpinia galangal*), dragon blood tree bark (*Dracaena loureiroi*), cardamom pods (*Amomum villosum*) for medicinal use; *No Kom* bamboo shoots (*Indosasa sinica),* elephant yams (*Amorphophallus paeoni-ifolius*), rattan stems (*Calamus* sp.), and bamboo worms (*Omphisa fuscidentalis*) for use as food and tiger grass flowers (*Thysanolaena latifolia*),[9] paper mulberry bark (*Broussonetia papyrifera*),[10] and *Puack Muack* (*Boehmeria* sp.)[11] for industrial materials. They also gathered small amounts of wild orchid species.

There were no differences in the species gathered by ethnic Lao and Khmu people. The NTFP with the greatest contribution to income was the bamboo worm, gathered by 13 ethnic Lao households and 14 Khmu households; one of the Khmu households earned an annual income of 6 million kip from bamboo worms alone. NTFPs in the village can be gathered by villagers from anywhere other than privately owned paddy fields. Households practicing swidden agriculture do not have ownership rights to secondary forests after rice harvest. The village permits village land to be used temporarily for swidden agriculture, but the land does not belong to the households who practice swidden agriculture. Therefore, even the NTPFs that can be gathered from secondary forests (e.g., tiger grass, paper mulberry, etc.) can be gathered by any villager and not just the household that practices swidden agriculture there. For landless Khmu people, the activity of gathering NTFPs anywhere in the village is the most important cash income activity, accounting for 50% of their cash income. There was a significant difference at a level of 5% between the Khmu people, who get a large proportion of their income from gathering NTFPs, and the ethnic Lao people, who do not.

5.3.5 Intermediating NTFPs

As for intermediating NTFPs, only three households (two ethnic Lao and one Khmu) were found, with no significant differences between ethnic groups. Intermediary sales of NTFPs in Laos exhibits stratification in the spatial range of transactions at the international and domestic levels (Yokoyama 2010). NTFPs intermediaries in HB village sell NTFPs gathered by villagers to intermediaries in the same area, such as in MN village and Nong Khiaw. They are intermediaries positioned at the remotest end of the chain among NTFP intermediaries.

5.3.6 Handicrafts

Fifteen ethnic Lao and six Khmu households received income from handicrafts. Weaving was the activity that generated the largest income, with about 58% of the village's handicraft income coming from weaving. All eight of the households practicing weaving were ethnic Lao; it is considered an extremely important economic activity for women of Tai ethnic groups in rural Laos (Kusakabe 2004). However,

---

[9] As raw material for making brooms, this is exported to Thailand, China, and Vietnam.

[10] This becomes the raw material for handmade paper; while some of it is used in domestic tourist sites in Luang Phabang and elsewhere, most is exported to Thailand.

[11] There is no English name for *Puack Muack*; it is a local Lao word. This material is used as glue or resin for securing sticks of incense. Depending on the region, it is also called *Sa-pan* or *Nang-nyao*.

since the Khmu of the Mon-Khmer ethnic group do not traditionally engage in weaving, no Khmu households in HB village were involved in this craft. Other handicraft items observed were brooms made with tiger grass and bamboo steamers and containers to be used for glutinous rice.

In the production of handicrafts, weaving is done by women and bamboo work by older people; this was especially common in ethnic Lao households that owned paddy fields. Therefore, a significant difference at a level of 5% was observed between the ethnic Lao, who had a lot of handicraft income, and the Khmu, who did not. There have been cases where handicrafts using bamboo resources were used in attempts at village development in farming villages in the capital area (Lee et al. 2021), but in HB village these were only sold within the village or in MN village, and the cash income was not much compared to that from other livelihoods.

5.3.7 Livestock

Livestock in HB village includes water buffalo, cattle, pigs, and poultry (i.e., chickens and ducks). Income was earned from the sale of any of these animals by 19 of 21 ethnic Lao and 6 of 17 Khmu households. There was a significant difference at a level of 1% between the ethnic Lao people who earned much income from animal sales and the Khmu people, who did not. This itemization is shown in Table 7. The ethnic Lao people raise larger numbers of all types of animals (e.g., water buffalo, cattle, pigs, and poultry) than the Khmu people, and they have greater household consumption and sales of the same.

Large livestock like water buffalo, cattle, and pigs are valuable as personal property (Millar and Photakoun 2008), and to rural households, livestock functions as a safety net for handling situations in which they can sell animals to cover large expenditures. Furthermore, livestock meat plays an important role as a source of protein for people and while they do not routinely butcher and eat large livestock, poultry often makes it to the dinner table. In the course of a year, 21 ethnic Lao households consumed or sold 397 poultry, but 17 Khmu households consumed or sold only 92 poultry. Clearly, there was a large difference between ethnic groups in the use of livestock.

5.3.8 Others

The other types of cash income classified in HB village included government allowances, remittances, tourism income, and the sale of wild game meat from hunting. However, there were no significant differences between ethnic groups. In HB village, there are many households with members working in urban areas. Although funds may, in practice, be flowing into households from these migrant workers, the money they earn is not remitted to households on a regular basis. It is often used all at once to purchase paddy fields or rebuild houses, and the present survey was unable to clarify this.

Table 7 Livestock raising and sales at HB villages, 2018

	Ethnic Lao (n = 21HHs)			Khmu (n = 17 HHs)		
	Total[(2)] (head)	Personal use or sales[(3)] (head)	Livestock sales[(3)] (Household)	Total[(2)] (head)	Personal use or sales[(3)] (head)	Livestock sales[(3)] (Household)
Water buffalo	54	12	5	8	0	0
Cow	35	14	7	14	2	1
Pig	79	25	13	28	5	4
Poultry[(1)]	357	397	9	126	92	2

Source Authors' interview survey
Note (1) Poultry includes chickens and ducks
(2) This is the number of animals raised at the time of the September 2019 survey.
(3) This is the number of personal use and sales in 2018.

5.4 Relationship Between Livelihood Structures and Ethnic Groups

The analysis of the ethnic groups and their forms of rice cultivation and cash income activities revealed that the majority of the ethnic Lao households in HB village held large areas of paddy fields and were able to produce enough rice to sustain themselves. However, the average rice harvest of the Khmu people, who immigrated from mountain areas in the 1970s and later produced rice with swidden agriculture instead of in paddy fields, is less than half of that of the ethnic Lao. While the ethnic Lao people gain income from selling surplus rice, the Khmu people who practice swidden upland rice cultivation generally do not produce a surplus.

Ethnic Lao households raise large numbers of livestock and have gained much income from selling these. The combined livelihoods of paddy farming and livestock raising are characteristic of the ethnic Lao. Meanwhile, Khmu households also raise livestock, but they have few large animals like water buffalo and cattle, and more poultry. In particular, large livestock such as water buffalo and cattle are raised by grazing in the mountains during the rainy season and in paddy fields after rice harvesting during the dry season. Therefore, it has been shown that it is difficult for the Khmu people who do not hold paddy fields to raise large livestock. Instead, the main cash income activity of the Khmu people was NTFPs gathering. NTFPs gathering can be done on the way to or while at swidden fields; therefore, it is highly suitable for swidden farmers who go to their fields almost every day during the rainy season.

6 Conclusion

Since the 1970s, residents from mountainous areas in northern Laos began to move into basin areas, and immigrants continue to be received in basin villages even today. In the basin villages that form the study area, which has constantly received migrants from mountainous areas, the ethnic Lao households who have been there longer practice self-sufficient paddy cultivation; by the 1970s, the land that could be reclaimed for paddy fields was already exhaustively developed. Nevertheless, the villages accepted migrants because there are highly productive paddy fields in the study area, allowing for rice subsistence in the region. However, on analyzing the status of paddy acquisition by immigrant residents at the household level, households that had wanted to engage in wet-rice cultivation and moved from mountainous areas to basin areas were found to be unable to reclaim paddy fields because of the lack of land suitable for paddy lots and, ultimately, practiced subsistence swidden agriculture even after migrating.

Furthermore, even within the basin area of the study area, settlements have been abandoned and migration has occurred, with residents of an abandoned settlement

(NM village) holding on to their paddy fields but moving to MN village, so the study area shows a notably complex situation of paddy holding.

An analysis of the livelihoods of the migrant Khmu, and the ethnic Lao households originally living in HB village revealed that ethnic Lao households that practiced paddy cultivation on fields inherited from their parents and raised livestock earned over twice as much cash income as Khmu households that practiced swidden agriculture and gathered NTFPs. However, even the ethnic Lao households that harvested enough rice to have a surplus still had cash incomes that were considerably lower than the average in Laos.

The relationship between paddy holding and population changes through migration from the mountainous villages to the basin villages and from the basin villages to urban areas is shown in Fig. 10. Not all migration from mountainous villages to basin villages is voluntary; in many cases, government initiatives aimed at inhibiting swidden agriculture, generating stable livelihoods, and consolidating small settlements led to migration. Meanwhile, migration from basin villages to urban areas happened voluntarily, as people sought a better lifestyle (i.e., education or medical care).

Even the ethnic Lao residents living for a long time in the basin of the study areas felt anxious about their future. Considering that agriculture is the only industry in these areas, it is difficult to expand paddy fields and get income from surplus rice; furthermore, infrastructure is lagging, there is no electricity, no roads are in place that cars can use, and there is not even cellphone coverage in these areas. Cash income is also low when compared to the average for Laos. Many residents voluntarily migrate

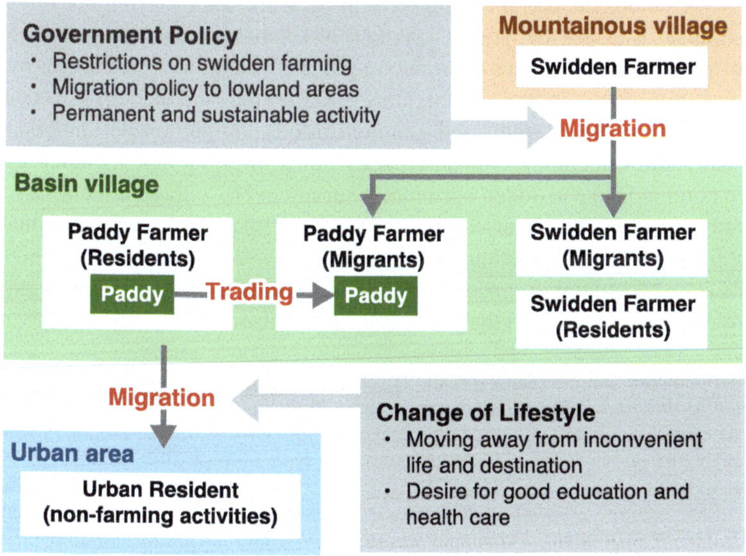

Fig. 10 Migration of mountain-basin-urban areas and acquisition of paddy fields

to cities to escape such an inconvenient lifestyle. Cash is needed to move to a city. Therefore, people sell their paddy fields and leave the village.

In the 48 years from 1970 to 2018, of the 73.22 ha of paddy fields in the study areas of NK and HB villages, 24.94 ha (equivalent to roughly 34% of the area) of paddy fields were being used after being purchased from residents who had moved to other villages. However, even if a paddy field is up for sale, households that have moved to the basin from the mountains or households of people who grew up in the study area but then got married and established an independent household separate from their parents do not have the funds to purchase paddy fields. These households that do not own paddy fields subsist on rice cultivated with swidden agriculture, while some of their household members go to work in urban areas to save money and wait for paddy fields to go up for sale. We also observed households that migrated from the mountains to the basin and then gave up on purchasing paddy fields and living in NK village or HB village to immediately migrate to urban areas. Évrard and Goudineau (2004) and Molland (2017) identified the high likelihood that resettlement compels further resettlement; in practice, the basin villages have become a temporary stopover for migration from mountainous areas to urban areas. In the present state of being unable to acquire paddy fields, residents who have migrated from mountainous areas are compelled to choose between staying in basin villages that do not have enough paddy fields and practicing swidden agriculture there or moving to urban areas (Niwa and Nishimoto 2021).

This study used long-term data on population change and data on paddy holdings related to the main livelihood of rice cultivation to reveal the instability of the circumstances in which the migrant Khmu people have been placed and the increasing tendency for ethnic Lao people, who have lived in the area for long, to migrate to cities because of a lack of land in small-scale Laotian communities, which have had constant and repeated migration due to political, economic, and social reasons. Topics for future research include obtaining longitudinal land data for swidden agriculture carried out by migrants and gaining a more comprehensive understanding of rice cultivation, including not only paddy cultivation but also swidden agriculture. We commenced a swidden agriculture survey in HB village in February 2019, but because of limitations caused by the COVID-19 pandemic, we were unable to conduct further surveys. A detailed analyses of the swidden agriculture practiced by migrants from mountainous areas is an important research topic that can be explored in future research, along with the possibility of acquiring paddy fields.

Acknowledgements The people of HK and HB villages graciously cooperated in the surveys as we conducted this study, and we are deeply grateful to them. We would also like to thank Mr. Sith PHAYBOUN, a former volunteer staff member of Ngoi District, Luang Pabang Province, for accompanying us on the survey. This study would not have been possible without Mr. Khamla SENGPHAXAIYALATH of the National Agriculture and Forestry Research Institute at the Laos Ministry of Agriculture and Forestry, who arranged the research trip, issued the research permits, and provided various on-site advices. This study was supported from JSPS Grants-in-Aid JP17H01633.

References

Baird IG, Shoemaker B (2007) Unsettling experiences: internal resettlement and international aid agencies in Laos. Dev Chang 38: 865–888. https://doi.org/10.1111/j.1467-7660.2007.00437.x
Chazée L (1999) The peoples of Laos: rural and ethnic diversities. White Lotus Press, Bangkok
Cole R (2022) Cashing in or driving development? Cross-border traders and maize contract farming in northeast Laos. J Agrar Chang 22:139–161. https://doi.org/10.1111/joac.12460
Committee for Planning and Investment, Lao Statistics Bureau (2012) Food security in Lao PDR: a trend analysis, Lao PDR expenditure and consumption survey (2002/03 and 2007/08 LECS). Lao Statistics Bureau, Vientiane
Évrard O, Baird IG (2017) The political ecology of upland/lowland relationships in Laos since 1975. In: Bouté V, Pholsena V (eds) Changing lives in Laos: society, politics and culture in a post-socialist state. NUS Press, Singapore, pp 165–191
Évrard O, Goudineau Y (2004) Planned resettlement, unexpected migrations and cultural trauma in Laos. Dev Chang 35: 937–962. https://doi.org/10.1111/j.1467-7660.2004.00387.x
FAO (1957) News of the world, Laos. Unasylva 11(1):42–43
Fox J, Fujita Y, Ngidang D, Peluso N, Potter L, Sakuntaladewi N, Sturgeon J, Thomas D (2009) Policies, political-economy, and swidden in Southeast Asia. Hum Ecol 37: 305–322. https://doi.org/10.1007/s10745-009-9240-7
Friis C, Nielsen JØ (2016) Small-scale land acquisitions, large-scale implications: exploring the case of Chinese banana investments in Northern Laos. Land Use Policy 57: 117–129. https://doi.org/10.1016/j.landusepol.2016.05.028
Friis C, Reenberg A, Heinimann A, Schönweger O (2016) Changing local land systems: implications of a Chinese rubber plantation in Nambak District, Lao PDR. Singap J Trop Geogr 37:25–42. https://doi.org/10.1111/sjtg.12137
Fullbrook D (2007) Contract farming in Lao PDR: cases and questions. Laos Extension for Agriculture Project (LEAP), National Agriculture and Forestry Extension Service (NAFES), Vientiane
Goudineau Y (ed) (1997a) Resettlement and social characteristics of new villages: basic needs for resettled communities in the Lao PDR, vol 1. United Nations Development Programme, Vientiane
Goudineau Y (ed) (1997b) Resettlement and social characteristics of new villages: basic needs for resettled communities in the Lao PDR, vol 2. United Nations Development Programme, Vientiane
Halpern J (1964) Economy and society of Laos: a brief survey. Monograph Series 5. Southeast Asia Studies, Yale University, New Haven
High H (2008) The implications of aspirations: reconsidering resettlement in Laos. CritAl Asian Stud 40:531–550. https://doi.org/10.1080/14672710802505257
High H, Baird IG, Barney K, Vandergeest P, Shoemaker B (2009) Internal resettlement in Laos: reading too much into aspirations: more explorations of the space between coerced and voluntary resettlement in Laos. CritAl Asian Stud 41:605–620. https://doi.org/10.1080/14672710903328039
Hua X, Zhang L, Kono Y (2022) Fruit booms and investor mobility along the China-Myanmar and China-Laos borders. Ecol Soc. https://doi.org/10.5751/ES-13380-270335
Ingxay P, Yokoyama S, Hirota I (2015a) Livelihood factors and household strategies for an unexpected climate event in upland northern Laos. J Mt Sci 12:483–500. https://doi.org/10.1007/s11629-013-2879-y
Ingxay P, Hirota I, Yokoyama S (2015b) Access to livelihood assets under the influence of a climate event in upland northern Laos. Geogr Rev Jpn Ser B 87:131–145. https://doi.org/10.4157/geogrevjapanb.87.131
Ireson CJ, Ireson WR (1991) Ethnicity and development in Laos. Asian Surv 31:920–937. https://doi.org/10.2307/2645064

Junquera V, Meyfroidt P, Sun Z, Latthachack P, Grêt-Regamey A (2020) From global drivers to local land-use change: understanding the northern Laos rubber boom. Environ Sci Policy 109:103–115. https://doi.org/10.1016/j.envsci.2020.04.013

Kallio MH, Hogarth NJ, Moeliono M, Brockhaus M, Cole R, Bong IW, Wong GY (2019) The colour of maize: visions of green growth and farmers perceptions in northern Laos. Land Use Policy 80:185–194. https://doi.org/10.1016/j.landusepol.2018.10.006

Kusakabe K (2004) Women's work and market hierarchies along the border of Lao PDR. Gend Place Cult 11:581–594. https://doi.org/10.1080/0966369042000307951

Lao Statistics Bureau (2020) Kansamruad khanxomsai lae khanxaychai khongkhuaheuan khangthi VI pi 2018/19 (The 6th Lao expenditure and consumption survey 2018/18). Lao Statistics Bureau, Vientiane (in Lao)

LeBar FM, Suddard A (eds) (1960) Laos: its people, its society, its culture. Hraf Press, New Haven

Lee B, Rhee H, Kim S, Lee JW, Koo S, Lee SJ, Alounsavath P, Kim YS (2021) Assessing sustainable bamboo-based income generation using a value chain approach: case study of Nongboua village in Lao PDR. Forests. https://doi.org/10.3390/f12020153

Lestrelin G (2011) Rethinking state–ethnic minority relations in Laos: internal resettlement, land reform and counter-territorialization. Polit Geogr 30:311–319. https://doi.org/10.1016/j.polgeo.2011.06.005

Lu JN (2017) Tapping into rubber: China's opium replacement program and rubber production in Laos. J Peasant Stud 44:726–747. https://doi.org/10.1080/03066150.2017.1314268

Manivong V, Cramb RA (2008) Economics of smallholder rubber expansion in Northern Laos. Agrofor Syst 74:113–125. https://doi.org/10.1007/s10457-008-9136-3

Millar J, Photakoun V (2008) Livestock development and poverty alleviation: revolution or evolution for upland livelihoods in Lao PDR? Int J Agric Sustain 6:89–102. https://doi.org/10.3763/ijas.2007.0335

Molland S (2017) Migration and mobility in Laos. In: Bouté V, Pholsena V (eds) Changing lives in Laos: society, politics and culture in a post-socialist state. NUS Press, Singapore, pp 327–349

Niwa T, Nishimoto F (2021) Internal migration in rural Laos: a case study of the village H in Luang Phabang Province. Jinkōgaku Kenkyū (J Popul Stud) 57:21–32 (in Japanese). https://doi.org/10.24454/jps.2102001

Openshaw K, Trethewie R (2006) Shifting cultivation, agroforestry, non-timber forest products and related livelihood systems in Laos: a case study. For Trees Livelihoods 16:359–380. https://doi.org/10.1080/14728028.2006.9752573

Phouyyavong K, Tomita S, Yokoyama S (2020) Smallholder's labor allocation for livelihood diversification: a case study in an upland village in northern Laos. Tropics 29:9–24. https://doi.org/10.3759/tropics.MS19-08

Thanichanon P, Schmidt-Vogt D, Messerli P, Heinimann A, Epprecht M (2013) Secondary forests and local livelihoods along a gradient of accessibility: a case study in northern Laos. Soc Nat Resour 26:1283–1299. https://doi.org/10.1080/08941920.2013.788429

Thongmanivong S, Fujita Y, Phanvilay K, Vongvisouk T (2009) Agrarian land use transformation in northern Laos: from swidden to rubber. Tōnan Ajia Kenkyū (Jpn J Southeast Asian Stud) 47:330–347. https://doi.org/10.20495/tak.47.3_330

Yokoyama S (2004) Forest, ethnicity and settlement in the mountainous area of Northern Laos. Tōnan Ajia Kenkyū (Jpn J Southeast Asian Stud) 42:132–156. https://doi.org/10.20495/tak.42.2_132

Yokoyama S (2010) The trading of agro-forest products and commodities in the northern mountainous region of Laos. Tōnan Ajia Kenkyū (Jpn J Southeast Asian Stud) 47:374–402

Yokoyama S, Takahashi S, Niwa T, Nishimoto F (2023) Population dynamics and paddy holdings in remote rural villages of Northern Laos. E-J Geo 18:291–308 (in Japanese). https://doi.org/10.4157/ejgeo.18.291

Rural-Rural Migration in a Rural Village, Luang Phabang Province, Northern Laos

Focusing Both on Immigration and Emigration in HB village

Takahito Niwa, Futoshi Nishimoto, Shinichi Takahashi, and Satoshi Yokoyama

Abstract This study revealed the demographic characteristics of rural regions of Laos from the perspective of rural–rural migration of the population focusing both on immigration and emigration in HB village, Luang Phabang Province. The results showed that both social changes associated with rural–rural migration, such as relocation policies and marriage migration, and natural changes after migration have a significant impact on population dynamics. In order to maximize the livelihood potential in destination villages, migrants consider the possibility of acquiring new farmland, the accessibility of farmland in their previous village, and the availability of kinship networks. The reason for active rural migration is that there is no significant change in livelihoods, mainly agriculture, before and after migration, and therefore migrants are able to continue supporting themselves. However, lowland rice paddies are limited. When it is difficult for relocated households to produce enough rice to sustain their families, they aim to earn cash by migrating to cities.

This chapter was published in Niwa et al. (2023).

T. Niwa (✉)
College of Human and Social Sciences, Kanazawa University, Kanazawa, Japan
e-mail: niwa-2@staff.kanazawa-u.ac.jp

F. Nishimoto
Embassy of Japan in Laos, Vientiane, Laos
e-mail: nsmtfts@gmail.com

S. Takahashi
Kobe University, Kobe, Japan
e-mail: takahashi@econ.kobe-u.ac.jp

S. Yokoyama
Graduate School of Environmental Studies, Nagoya University, Nagoya, Japan
e-mail: yokoyama.satoshi.i4@f.mail.nagoya-u.ac.jp

S. Yokoyama et al. (eds.), *Population Dynamics and Livelihood Changes of Small-Scale Societies in Laos*, International Perspectives in Geography 22,
https://doi.org/10.1007/978-981-96-2956-5_8

Keywords Population migration · Rural-rural migration · Population dynamics · Laos

1 Introduction

Laotian farmers have been actively migrating for a long time (Évrard and Baird 2017), however, it would be difficult to say that the features of this migration concerning the regional populations in rural areas of Laos are fully understood. Furthermore, the research on regional demographic changes in Laos is limited. This is due to the lack of reliable statistical data in Laos, making quantitative analysis difficult (Rigg 2007; Hickey et al. 2013). Moreover, economic growth in urban areas such as Vientiane, Savannakhet, and Luang Phabang is attracting migration from rural areas (Niwa and Nishimoto 2025). However, migration is happening actively not only between rural and urban areas but between different rural areas as well. This chapter will use the case example of a farming village in Luang Phabang Province located by the northern border of Laos to discuss the characteristics of rural–rural migration in farming village demographic changes, a topic that has received little previous attention.

Phouxay (2010) summarized an overview of migration in Laos, highlighting socioeconomic aspects while segmenting the 25 years since the founding of the nation of Laos into four phases to study the characteristics of migration in Laos. The first phase is 1975–1980, when the civil war ended and the nation was founded as the Lao People's Democratic Republic, and a socialist government and social system were constructed. In this period, migration was primarily that of citizens returning after having escaped during the war and refugees migrating outside the country. The second phase is 1985–1990, around the 1986 announcement of "new economic mechanisms," which triggered the introduction of the market economy. Concurrent with the return of refugees, there was more rural–urban migration caused by the economic reforms in urban areas. The third phase, the period from 1995 to the 2000s, saw the development of networks in the country, and simultaneously, the advancement of initiatives by the government and international organizations to develop farming villages and reduce poverty. In this phase, in addition to rural–urban migration, rural–rural migration increased, particularly from high ground to the lowlands. The fourth phase corresponds to international economic progress from 2005 onwards. In addition to domestic migration, international migration has gained attention. To understand the current migration in Laos, one needs to focus on three points: rural–urban migration, rural–rural migration, and international migration.

Of these, rural–urban migration has been discussed in the context of employment opportunities, education opportunities, and infrastructure maintenance concentrated in Vientiane (Phouxay and Tollefsen 2011; Insisienmay and Philavanh 2018). International migration, especially migration to Thailand, has been discussed in terms of the context of the Thai government's development of immigration policies and the high linguistic and cultural affinity between Thailand and Laos. It has been positioned as a link in household livelihood strategies based on economic motives and discussed in terms of family and local networks supporting chain migration (Skeldon 2006; Manivong et al. 2014; Niwa et al. 2025).

Rural–rural migration, in comparison, has been studied with a focus on a particular point of view, the migration related to the Lao government's resettlement policy. The government resettlement policy is a policy for having minority ethnic groups living in the highlands migrate to lowland areas.[1] This policy prohibits swidden agriculture and opium cultivation that was practiced in the highlands and encourages the cultivation of cash crops based on plantations, thereby converting traditional farm village economies into a modernized market-oriented system (Bird 2009). At the same time, the policy aims to improve access to government services for people in rural areas, including education, health and hygiene, agricultural services, electricity, and road infrastructure.

Much of the government resettlement policy is not conducted with villagers at its core, but they are not openly coerced by the government. Nevertheless, in practice, several instances have been pointed out regarding the vulnerability of minorities and women due to policy coercion and migration. For example, Baird and Shoemaker (2007) note that due to the government resettlement policy, minority ethnic groups are culturally and economically pressured to adapt to the dominant ethnic Lao lifestyle. In discussing village variations of swidden agriculture, Nakatsuji (2013) points out the same issue between ethnic groups that Baird and Shoemaker (2007) mention. A survey conducted in multiple villages in Xieng Ngeun District, Luang Phabang Province, found that the Khmu and Hmong people who had lived in the highlands and practiced swidden agriculture were migrating to "key area" target villages alongside main roads in lowland areas starting from the 2000s. As it is difficult to continue engaging in swidden agriculture in lowland villages with high population densities, they are diversifying their livelihoods, such as by working as day laborers. However, Nakatsuji (2013) mentioned:

> those who work as day laborers are Khmu in the poorest population segment, and most of the employers are wealthy people in Thai ethnic groups. Quitting swidden and reinforcing the inclination toward jobs like these will end in reinforcing economic domination–subordination between ethnic groups (Nakatsuji 2013: 349).

Previous studies on rural–rural migration focused on the people forced to migrate due to the government resettlement policy, or on the damage that the policy wrought on their livelihood. Conversely, the fact that this migration is occurring could be understood as the higher pressure of population having a large impact on the rural areas that receive the migration. However, there is insufficient understanding of

[1] Initiatives are being advanced through the Focal Site Development Project that gathers ethnic minorities into designated areas, the Village Consolidation Project that consolidates small villages and concentrates people into areas with high population density, and the Land and Forest Allocation Project that assigns village borders and allocates farmland to each household to manage natural resources (Baird and Shoemaker 2007). Ducourtieux et al. (2005) have pointed out that the land allocation has a negative effect on forest protection and agriculture.

the population changes and migrations in rural areas, which would be necessary to understand the socioeconomic spatial structure of Laos.

Thus, this chapter aims to clarify the characteristics of demographic changes in rural Laos from the aspect of migration. We especially focus on rural areas where migration policies have been implemented to promote in-migration, and where out-migration to urban areas and other rural areas has been active.

2 Research Methods

This study uses the case of HB village in Luang Phabang Province to examine the features of rural–rural migration in Laos (Fig. 1). We start by clarifying the features of incoming and outgoing migration in HB village, based on the demographic changes reconstructed from surveys on all households in HB village and interviews with the village head and elders. The demographic changes used in the analysis included family structural changes, such as births, deaths, marriages, and incoming and outgoing movements, as well as information tracing and reconstructing the deaths and outgoing movements of people who used to be registered in the village (Nishimoto 2011; Niwa and Nishimoto 2025).[2] In the case of HB village, we recorded 463 people and were able to trace back to the pioneers who founded the village at the start of the twentieth century. These research methods produced valuable data in rural Laos, which does not have detailed demographic data. However, there was a tendency to estimate the population as being lower the further we traced to the past as there was forced migration and evacuation due to the civil war in Laos until the early 1970s. Therefore, we only present data starting from 1975 in this chapter.

Next, to reveal the characteristics of the migration of HB villagers, we conducted structured interviews with 15 people who had migrated out of HB village to elsewhere in Luang Phabang. The survey period was August 20–23, 2019. We then conducted an additional survey in HB village from March 16–19, 2020, featuring semi-structured interviews with five households that had migrated to HB village. Each survey was conducted in the presence of counterparts in the Lao National Agriculture and Forestry Research and the Lao Tropical and Public Health Institutes.

---

[2] Specifically, we used the retrospective method. We determined the age of the current villagers and confirmed their place of birth. Then, we clarify the chronological changes in the members of each household by recording the year of marriage, year of migration, destination. Through interviewing to the village head and elders, we determined the number of people who have already left HB village or died. However, the further back in time, the more ambiguous the villagers' memories become.

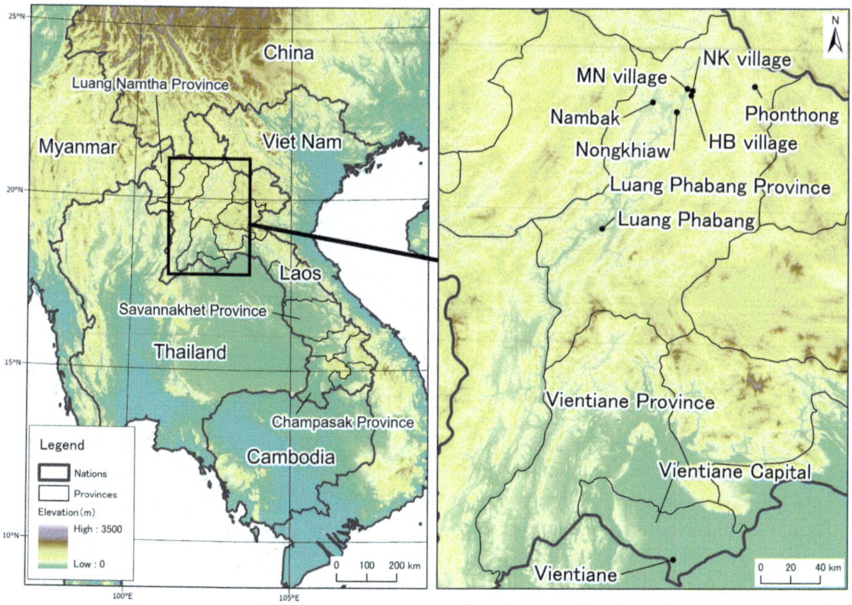

Fig. 1 Location of the field site in Luang Phabang Province, Northern Laos. *Note* Vientiane Capital is one prefecture in Laos. Vientiane is the city center

3 HB Village Demographic Changes and Migration

3.1 HB Village Population

The population of HB village as of 2018 is 208. Note that here we include those that have temporarily migrated in the villager population, the details of which will be mentioned later. The population includes 96 males and 112 females. The ethnic composition is 91 ethnic Lao, 115 Khmu, and 2 people of other ethnicities. There are 43 households,[3] of which 24 are ethnic Lao and 19 are Khmu.

Since 1975, except at certain times, the population has shown an upward trend (Fig. 2). There has been a slight decrease from 211 people in 2016; however, since the 2010s, the population has almost leveled off and stabilized. The proportion of ethnic Lao people has fallen from about 75% in the 1970s to 44% in 2018. The village had been majority ethnic Lao, but now there is a greater degree of different ethnicities living together. The level of ethnic mixing was already noteworthy as of 1975.[4]

[3] A common household is composed of a married couple or parents and children (Dommaraju and Tan 2014). It is difficult to accurately determine the ethnic identity of children when a couple has different ethnic groups (Nakata 2004). In this chapter, the ethnicity of the children was uniformly matched with that of their fathers.

[4] There are occupational and cultural differences between the ethnic groups (Nakatsuji 2013). In neighboring NK village, annual events run separately by each ethnic group began to be attended by

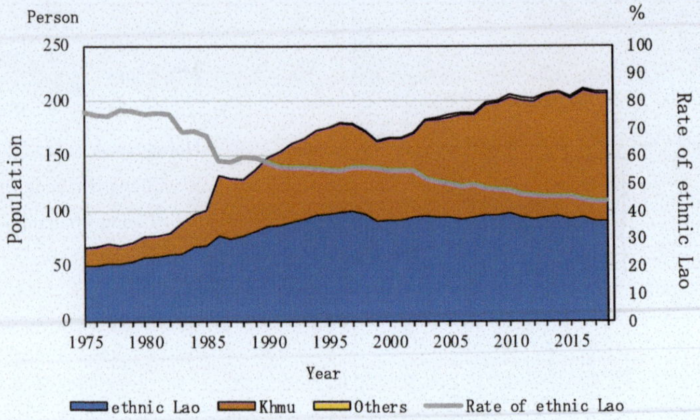

Fig. 2 Population change by ethnicity in HB village, 1975–2018. *Source* Authors' interview survey

Considering the demographic changes for the ethnicities of Lao and Khmu at five-year intervals (Fig. 3), both increased over the past 50 years, but they are trending downward; both groups approached a level of natural increase to around zero in 2015–2018. We would need to investigate the child mortality and birth rates by women's age groups for details, however, the number of births is decreasing due to the spread of family planning.

Meanwhile, social increase has trended downward over the long term. Additionally, in 2015–2018, there was an excessive outflow of ethnic Lao and Khmu. Comparing ethnic Lao and Khmu, the size of inflow of Khmu in the 1980s is striking. Regarding ethnic Lao social increase, the number of outflows was already bigger than that of inflows by the 1990s. On the other hand, the Khmu only experienced social decrease in 1995–1999 and 2015–2018, having excessive inflow in the other periods. This is due to immigration from other villages. Furthermore, after ethnic Lao and Khmu experienced excessive inflow in the 1980s, the natural increase reached a peak in 1990–1994 because families grew after moving to HB village.

In the following subsects. 2 and 3, we will show the characteristics of in-migration and out-migration, which have been a major factor in the demographic changes of HB village.

---

both ethnic groups in 2015, upon the proposal of the former village head (according to an interview with the current village head, conducted in March 2020).

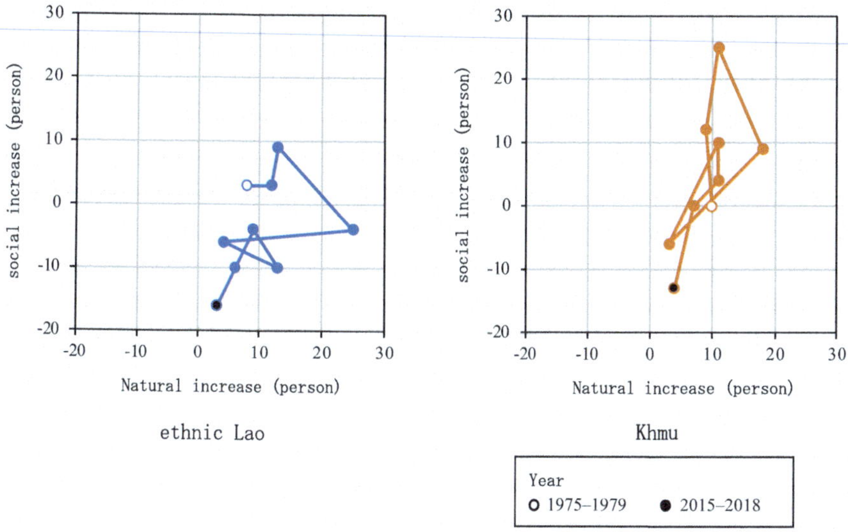

Fig. 3 Population change as shown by natural change and net migration in each 5-year period, 1975–2018. *Note* 2015–2018 is not a 5-year interval. The lines indicate that they are in a time series. *Source* Authors' interview survey

3.2 In-Migration to HB Village

3.2.1 Overview of In-Migration

From 1975 to 2017, 140 people moved into HB village. These are further classified into 48 single people and 31 households, which included 92 people. Inflow occurred almost every year, but in certain years the number increased. The years with an inflow scale of more than ten people were 10 in 1983, 31 in 1986, 11 in 2003, and 13 in 2013. These four years account for about half of all the in-migration during 1975–2017. In 1986, the year with the greatest inflow, 7 of the immigrants were single and the other 24 moved in as households. Furthermore, the birthplaces of 31 of these people was concentrated in three areas, with 28 people coming from one area, Phonthong District (Fig. 1). The 13 immigrants in 2013 comprised 2 households and 1 single person. Household migration was the reason for the size of excessive inflow. This rural–rural migration is in large part because of the migration policy.

The migration policy is a major factor of rural–rural migration in Laos, however, there are other types of rural–rural migration. One mainstream pattern is marriage migration; classifying cases where the migration and marriage years were the same as marriage migration, this would apply to 21 of the 140 people. Twelve women and nine men had marriage as a reason for immigration, with no significant difference between genders.

Of these 140 people, only 75 are living in HB village as of 2018. From the other 65, 30 have died, 6 have migrated elsewhere temporarily, and 29 have migrated elsewhere permanently.

Furthermore, of the 43 households, only five households had no members from outside the village; in other words, all household members were born in HB village. These households were marriages between people from the village, and of these, four were ethnic Lao. There were many households with only one person born outside of HB village, a total of thirteen households. In these cases, one partner in the marriage moved in from outside the village. In contrast, there are 11 households where more than half of the members were born outside the village; of these, 4 had no members born in HB village.[5] These households migrated as a household or were marriages between people born in other villages.

3.2.2 Examples of Immigrant Household

Here we will introduce one case each of incoming ethnic Lao and Khmu households, and give an overview of the state of household migration.

Ethnic Lao Household A moved to the village from within the province in 2014, with seven members; the head of household was 49 years old. In addition to the household head couple, there were two children, one spouse of a child, and two grandchildren. Their reason for moving in was the government resettlement policy. When they moved in, they consulted with the village head, were given a house by the village, and purchased agricultural land in the village. Currently, one of the child households from Household A that had grandchildren has built a new house and become independent. The other child married a woman from another village in 2017, and lives with Household A. For this reason, as of 2018, Household A has four members. They engage in wet-rice cultivation in the village and do not engage in swidden agriculture. Their rice is not enough to meet the demand of the household. Therefore, they purchase the rest. Meanwhile, they possess many heads of livestock, such as buffalo, pigs, ducks, and chickens. Their profits on sales of livestock have an important position in their livelihood and in to purchase rice.

Khmu Household B moved to the village in 2008. There were a total of seven members, including the household head couple, three children, and the parents of the head of household. The younger daughter had moved to HB village ahead of them when she married in 1992. She consulted with the village head ahead of time before her father, the head of household, asked the head for permission to move to the village. Household B stated that their reason for moving was that there was not enough land in the former village they lived. This is positioned as migration seeking a better life. At present, they engage in swidden agriculture in the vicinity of HB village and do not have paddy fields. They purchase the rice that they need. They do not have any livestock, and their meager family income is gained from non-timber

[5] However, it should be noted that the more time has passed since moving in, the more members of the household were born in HB village.

forest products. Additionally, one of the daughters has gone to work in Vientiane; her remittances form an important source of household income.

3.3 *Trends in Permanent Out-Migration of HB Village*

3.3.1 Overview of Out-Migration

Next, we will confirm the outflow from HB village to other areas. The population of HB village is 208; at the time of the survey, 25 of them were not residing in the village. They are counted in the village population because they are considered to have left the village temporarily. They are people who have left the village for work or education and are temporary migrants who may return to the village in the future. However, there are many permanent migrants who are unlikely to return to the village.[6]

Despite that Niwa and Nishimoto (2025) have analyzed temporary migrants from HB village, we analyze the 99 permanent migrants of HB village in this chapter.[7] First, we will look at migration in three types, based on the destination of the move: rural–urban migration, rural–rural migration, and migration to other provinces or countries. We, therefore, positioned only some destination places such as Vientiane, Luang Phabang, and Vientiane Province as rural–urban migration because information beyond the district or provincial level was unavailable for destinations in some cases. As a move to another province could be rural–urban or rural–rural migration, these were treated other types. Aside from these, the moves mainly within Luang Phabang Province were defined as rural–rural migration.

There was a total of 51 rural–urban migrations, with 37 to Vientiane, 7 to Vientiane Province, and 7 to Luang Phabang. About half the permanent migrations headed for cities and are positioned as mainstream migration. There were 15 people who migrated to other provinces or countries, with a variety of destinations, from as near as neighboring Luang Namtha Province to as far as Savannakhet Province. One person migrated to another country, Vietnam. It was not possible to determine from the interviews with villagers whether the migration to other provinces and countries was rural–urban or rural–rural migration. The remaining 32 people can be classified as having engaged in rural–rural migration. These include one person who moved to the neighboring NK village and six to MN village, six others moved to Nam Bak District, located midway between the village and Luang Phabang.

Next, we will examine the occupations of the migrants at these destinations. However, as this is the information that the villagers of HB village had, it may not

[6] Temporary migrants refer to people who leave home to study or work in an urban area. In contrast, permanent migrants are positioned as people who have left with their family, such as couples or parents and children. This classification follows the description of the head of HB village.

[7] We only picked up out-migrants who left the village permanently after 1975 and were considered alive as of 2018.

Table 1 Working status of permanent emigrants by migration type

Migration type	Primary industry	Secondary industry Tertiary industry	Student	Total
Rural–urban	14	30	6	50
Rural–rural	22	5	3	30
Others	3	6	0	9
Total	39	41	9	89

Source Authors' interview survey

be the most recent information. Even so, knowing the kind of job the migrants had at the time of the move could help us understand the context of migration. While there are some clear examples, such as civil servants (8), soldiers (2), and farmers, there were cases where the person was simply engaged in some kind of employed labor (28), and others where the occupation was completely unknown. We divided the employed labor whose industry was unknown into primary, secondary, and tertiary industries based on data compiled by Niwa and Nishimoto (2025).[8] Excluding 10 people whose occupation was unknown, the 89 remaining migrants were sorted into 39 in primary industries, 41 in secondary or tertiary industries, and 9 who were not working and were students (Table 1).

Comparing these with the type of destination of the migrants who headed for urban areas, 30 were engaged in secondary or tertiary industries. Meanwhile, 22 of the rural–rural migrants were engaged in agriculture. There is a big difference between the employment content of rural–urban migration and rural–rural migration. Of those who were students, six moved to urban and three to rural areas. The proportion of people migrating for education is not small.

Reviewing the relationship between the years in which permanent migrants moved and their age (Fig. 4), we can confirm that there has been migration outside the village for a long period, regardless of whether the destination was urban or rural. Despite the move period, there are many people around the age of 20 who move. However, since the 2000s, migration has expanded to a broader expanse of age groups.[9] Comparing the age at the time of the move and the type of migration (Table 2), the average ages for rural–urban and rural–rural migration was 21.9 and 21.5, respectively, showing no major difference. Adding in variations in age, there was no difference in the timing of the move between rural–urban and rural–rural migration. As there were older people of 60 or above migrating to other provinces or countries, the average age was high; there was also a large standard deviation.

[8] Employed labor includes working in plantation farming or working as a tenant farmer. As per Niwa and Nishimoto (2025), in Vientiane, of the 12 people in HB village who were said to have worked in employed labor, 4 were working in agriculture. Thus, in this chapter, we proportionally distributed the 25 people who were employed later in rural–urban migration according to this ratio, with 8 in primary industries and 13 in secondary or tertiary industries.

[9] Out-migration has presumably been active since the 2000s, however, we cannot negate the possibility of data omissions for past emigrants in the 1980s and other periods.

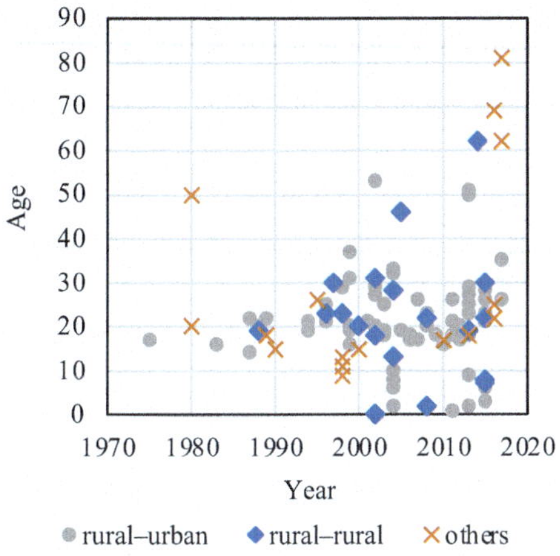

Fig. 4 Migration year and age of permanent emigrants by migration type. *Source* Authors' interview survey

Table 2 Migration age of permanent emigrants by migration type

Migration type	Average age	Standard deviation
Rural–urban	21.9	11.3
Rural–rural	21.5	11.2
Others	29.4	22.0
Total	23.0	13.9

Source Authors' interview survey

The context for the distribution of movers' ages includes not only migration based on individual decision-making but also decision-making at the household level, for reasons of household migration or livelihood strategy. Therefore, we will analyze the characteristics of out-migration at the household level.

Comparing temporary and permanent migrants at the household level, only permanent migrants who are members of existing households are shown, a total of 50 persons. Table 3 shows the number of village residents, temporary migrants, and permanent migrants per household by ethnic group. There was no difference between ethnic Lao and Khmu in the average number of household members, for the number of actual village residents (4.4) or with the addition of temporary migrants (0.6). The average household has 5.0 people. However, there was a difference in the number of permanent migrants who moved outside the village from the household, with an average of 1.7 for ethnic Lao and 0.6 for Khmu. The difference is not statistically significant.[10] This is probably because the length of time of household formation also affects the number of permanent migrants.

Figure 5 shows the number of people who moved as a proportion of household size in relation to the year of marriage of the head of household, and is separated

[10] Since the number of persons per household cannot be considered normally distributed, a nonparametric test, the Wilcoxon signed rank sum test, was performed ($p = .2096$).

Table 3 Average number of household members by migration status and household ethnicity

unit: person

Household ethnicity	Residents	Temporary migrants	Permanent migrants
Ethnic Lao (n = 23)	4.4	0.5	1.7
Khmu (n = 19)	4.3	0.7	0.6
Total	4.4	0.6	1.2

Source Authors' interview survey

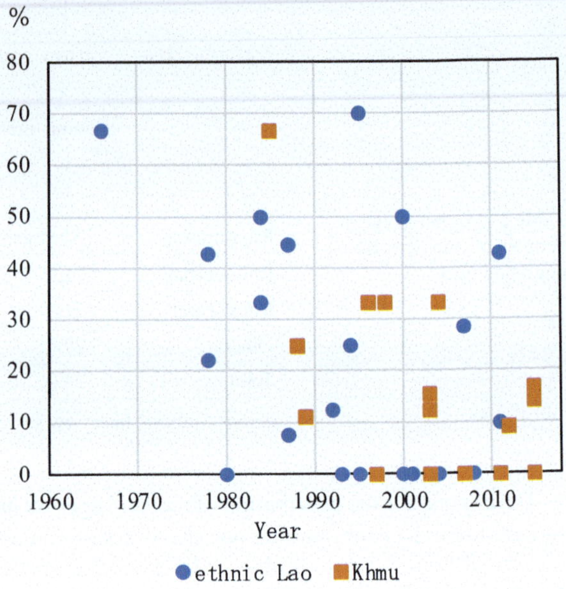

Fig. 5 Marriage year of the head of household and ratio of emigrants in the household by household ethnicity. *Note* Liner regression equation and coefficient of determination are respectively y = −0.672 x + 1384.3 and $R^2 = 0.21$. *Source* Authors' interview survey

by ethnic group. The out-migrants include temporary and permanent migrants. The proportion of migrants in a household tends to grow when more time has passed since the marriage. This is similar for ethnic Lao and Khmu, but the slope of the first-order regression is slightly higher for the Khmu. This figure simply shows the relationship between the year of marriage and the proportion of household members who have out-migrated. It cannot be asserted that there is a relationship between them. We will explain this point in detail in the next subsection.

3.3.2 Specific Migration Examples

Here will introduce the cases of five permanent migrants. All of them live in Luang Phabang Province; they live in different places and have different life circumstances. There is a pattern of having livelihoods that differ from the occupations of HB village the farther away from the village they are. Additionally, we will introduce two cases of siblings moving to Luang Phabang, as an axis of comparison between rural–rural and rural–urban migration.

Person A is an ethnic Lao woman who was born in 1938 and had 14 children.[11] She left HB village in 1958 when she got married and lived in neighboring NK village where her husband's hometown. Then, in search of education opportunities for her children, the family of ten moved to MN village in 1993 as they could provide their children with up to secondary level education there. Her children moved to Luang Phabang and other places, and her household currently consists of five people, including one daughter and son-in-law, and two grandchildren. The biggest reason for their living near HB village is access to farmland.[12] As their current livelihood is based on agriculture, they purchased farmland in MN village before they moved and have continued to engage in wet-rice cultivation. They also raise over ten heads of cattle. Her daughter and son-in-law have started a tourist business. Further, she has secured 0.95 ha of farmland in HB village. She does not use it herself and instead rents it to her nephew in HB village.

Person B is a Khmu woman who moved to Nong Khiaw in 2001 when she got married. Nong Khiaw is about one hour down the river from MN village, and is a central town with the government office of the district that contains HB village. She currently lives with her husband, his parents, and their four children in a household. Person B was born in 1982, after finishing five years of elementary education, she went to work in Vientiane from 1999 to 2000. This happened because someone from MN village was gathering people to work at a garment factory. She met her husband while on a short trip home and decided not to return to Vientiane. Since marrying she has lived in Nong Khiaw and currently works in agriculture. Her husband migrated to Nong Khiaw because of the government relocation policy but still has farmland in his previous village. In addition to agriculture, he operates a tourist business in Nong Khiaw. Person B has already turned over her rights to use HB village farmland to her siblings and does not use it herself. For this reason, her relationship with HB village is getting thinner.

Person C is an ethnic Lao born in 1982 in HB village, who studied up to the third year of secondary education in MN village before migrating to Vientiane to follow his older brother who was employed there. Person C went to school in Vientiane for one year, and then from 1998 continued his studies at a military school, introduced by his older sister, in Champasak Province. It was perhaps around this time that he switched from temporary to permanent migration. After finishing his secondary education, he enlisted in the army and continued his studies up to the bachelor's level in 2010. After graduation, he was stationed in a unit in Nam Bak District and continues to stay there. In 2016, he married a person who was also in the army. His parents still own farmland in HB village, but he does not intend to inherit it. In the future, he plans to pass this on to his siblings who have been staying in the village.

[11] The 14 children are based on the interview survey with Person A. The number of children is extremely large, even if it was a period of high fertility. The population data based on the survey in HB village does not record her children.

[12] Other migrants living in MN village also engage in farming.

Person D is an ethnic Lao born in 1992, who studied in MN village up to the third year of secondary education. In 2013, he went to work in a factory in the suburb of Luang Phabang. His father saw a recruitment poster for a food processing factory posted in MN village, and after discussing it as a family, they decided that Person D would go to work there. He was employed there until 2016 when his father told him to come back to help with farming, and he returned to HB village for six months. In 2017, he migrated again to the suburb of Luang Phabang, and he currently works as a driver for another factory. He switched from temporary migration to permanent migration when he married in 2017. He lives with his wife and her parents in their house with their one child. His wife has started to raise a small number of pigs while this has produced no income. Person D's income is the source of household income. Further, he sends a yearly remittance to his parents in HB village from his income.

Person E is Person D's younger sister, born in 2002. She currently lives in the home of the parents of her fiancé in Luang Phabang. The household has seven people, including her fiancé, his parents, and the three-person family of her fiancé's younger sister. Person E went to work in Luang Phabang in 2018 after a discussion with her parents ahead of migration. She met her fiancé when she worked for about one month as a waitress at a restaurant. Her fiancé is a soldier in the army; his father was also a soldier. They are currently in the process of getting married and are during the procedures for changing the family register at the district office. During the interview, she was in the process of switching from temporary to permanent migration. As the procedures are not yet complete, she has not married yet; because of the process, she cannot currently work. She thinks she would like to work again in the future. Persons D and E do not have farmland they use themselves in HB village, however, their parents are in good health and continue to use the farmland in HB village. In the future, even if they inherit the rights to the farmland, they have only a faint notion of going back to the village and engaging in farming. This is because they have family in Luang Phabang and have transferred the basis of their livelihoods there. Both intend to pass the land to their siblings who stayed in the village.

4 Discussion

The migration occurring in HB village plainly shows the changes in the local population that is happening in lowland villages, in the context of the Lao government's resettlement policy. HB village was majority ethnic Lao and had only a small number of Khmu in 1975, but the village's ethnic makeup has had a reversal for two reasons. The growth of the Khmu population is due to their moving in as households starting from the 1980s and their high birth rate. In contrast, the stagnation or slight decrease in the ethnic Lao population is due to the back-and-forth between their natural increase and social decrease. In other words, natural and social increase have influenced the population structure of HB village. This can be understood by the proportion of people born in the village by household. Regardless of ethnic group, the more time that passes since the household was formed, the more children are born in the village.

Therefore, the proportion of the household born in the village increases. As this is influenced by the aging of the household, the fact that households with a smaller proportion of members born in the village being in the minority means that there are many households that have recently moved to HB village.

Furthermore, as seen in the incoming moves to HB village, there has been movement not only at the household level but also at the individual level. Of these, moving because of marriage has been an important factor. Generally, ethnic Lao husbands live in the homes of their wives, while Khmu wives enter the homes of their husbands (Thomson and Baden 1993). One feature of rural–rural migration is that it occurs regardless of gender.[13]

The government resettlement policy has influenced in-migration, causing great changes in the population structure of HB village. This does not occur only among ethnic minority households such as the Khmu but is also seen among the ethnic Lao (See case of Household A). Additionally, people who migrate under this policy do not migrate to places designated by the government, they must search for their destinations themselves. Although the receiving villages must accept the moves of new migration applicants, the migrants do not choose their destinations blindly. They consider the possibility of acquiring new farmland, accessibility to farmland in the previous village, or the presence of a network of family and local relationships to maximize the livelihood potential of their destination (See the cases of Household A and Household B).

This perspective is an important element of rural–rural migration outside of the government resettlement policy. The case of Household B, which moved to HB village, suggests that as basic infrastructure was not available in their previous village, the only way access to better infrastructure was to migrate. This can be seen not only in in-migrations to HB village but also in out-migration from the village (See case of Person C). Even in the case of rural–rural migration, it is possible that the direction of outward migration is directed toward villages with a more urban character. Children from HB village are sent to the neighboring MN village from the level of elementary education; for upper secondary education, they need to be sent at least as far away as Nong Khiaw (Fig. 1). Access to more complete basic infrastructure, such as education, employment opportunities, and the benefits of infrastructure development, materializes in the awareness of migrants.

In other words, rural–rural migration in Laos, especially in northern Laos, can be thought of as a way of coping with regional disparities in infrastructure and socioeconomic conditions.[14] The government resettlement policy in rural areas of

[13] There are also Lao and Khmu couples. There are 13 Lao-Khmu married couples among the couples living in or temporarily away from the village. Among them, 6 couples have Lao husbands, and there is no significant difference between the two ethnic groups. Therefore, inter-ethnic marriages have not affected the ethnic composition of the village currently.

[14] The case studies in this chapter suggest that regional socioeconomic disparities have improved somewhat as a result of population migration. On the other hand, there is a strong argument that population migration is not the optimal solution for correcting regional disparities, as policies to promote migration have exacerbated poverty (High 2008) and widened social disparities in the transition to commercial agriculture (Rigg and Wittayapak 2009, Andriesse and Phommalath 2012).

northern Laos has had a significant impact on areas on the boundary between highland and lowland areas, or where both swidden and paddy fields coexist. In HB village, the ethnic Lao are more proactive than the Khmu about improving accessibility to basic infrastructure through outward migration. This is seen in the fact that they began to migrate out of the village at an earlier stage and that more of their household members became permanent migrants. Emigrants' connections to the village weaken over time; they switch from temporary to permanent migration. This is why the average number of permanent migrants from ethnic Lao households is 2.6 times greater than from Khmu households. While it is highly likely that this is affected by the age of each household, we can understand that the ethnic Lao move in search of a better life, and the Khmu move into vacated lowland villages in search of new places of residence.

Compared with rural–urban migration in search of wages, in rural–rural migration, many people continue working in agriculture even after migrating and are unlikely to migrate again for work (See the cases of Person A and Person B). The continuation of agriculture in rural–rural migration can be perceived from two aspects. First, they can use the knowledge and skills they learned before migration even after they migrate. In continuing to farm, migrants' adaptability is high if there are no major changes in the farming environment, such as climate (Bazzi et al. 2016). They can maintain their household livelihood by continuing to work in agriculture after migrating. Second, in rural–rural migration, there are no employment options other than farming, as job opportunities are limited.

Meanwhile, households that do not have farmland usage rights are unable to meet their food needs and have no choice but to purchase their staple food rice. In these cases, migration in search of employment opportunities is not permanent rural–rural migration; instead, it is characterized by temporary migration that maintains ties with the village (See case of Person D and Person E). When a household comes to be affected by the market economy and as a result, the children leave in search of work, we can position this as part of that household's livelihood strategy; in such a scenario, non-farming income takes on greater importance (Cole and Rigg 2019). It is also conceivable that cooperativeness surrounding survival is changing within villages. Previously, a generally accepted rule was that farmers could make a living even if they didn't have self-owned farmland, and they could borrow rice from relatives for their survival even if they were unable to harvest enough (Rigg 2019: 49–50). This safety net is thought to have weakened, however, and livelihood strategies at the household level have become more important.

5 Conclusion

In this chapter, we clarified the characteristics of rural–rural migration in rural demographic changes in the case of HB village, Luang Phabang Province, northern Laos, through reconstructed demographic data and interviews with migrants. The results showed the following three effects of rural–rural migration on population changes in farming villages.

(1) Social increase in the context of rural–rural migration due to government resettlement policy and marriage migration, and natural increase following migration have a significant impact on population change in rural Laos. In rural–rural migration, people consider the possibility of acquiring new farmland, accessibility to farmland in the previous village, or the presence of their family and local network to maximize the livelihood potential of their destination.
(2) Rural–rural migration in northern Laos can be thought of to cope with regional disparities in infrastructure and socioeconomic conditions. A chain reaction of migrations in search of a better life has resulted in a reversal of village ethnic composition in the case of HB village.
(3) In the context of active rural–rural migration, it is likely that there will be no great change in the core occupation of agriculture before and after migration, and that livelihoods can be maintained. However, available lowland paddy fields are limited; when incoming households have difficulties obtaining enough rice to maintain their livelihood, they aim to acquire cash by working away from home.

The characteristics of population migration suggest that governmental relocation policies have triggered not only migration from highlands to lowlands but also migration between lowlands. This may also affect the demographic dynamics of rural areas, such as ethnic mixing and changes in the ethnic composition of the population. These complex migration processes have the possibility to be discussed as characteristics of rural–rural migration in the rural areas of northern Laos, which is considerably different from lowland paddy field rural areas such as central Laos (Takahashi et al. 2025). It is necessary to collect data not only from the surrounding areas of HB village but also from other provinces to verify the findings. Population structures in rural Laos are affected by rural–rural migration and changed by it; it is possible that household-level livelihood strategies are becoming more important in response to this. Although the relationship between population and occupation has been demonstrated by Yokoyama et al. (2025), there has not been enough discussion on the aspects of births and deaths on the population structure. This will be a future topic of discussion.

Acknowledgements This study was supported from JSPS Grants-in-Aid JP17H01633. We would like to thank everyone involved who cooperated with the field survey. We would also appreciate the assistance of Dr. Khamla Sengphaxaiyalarh in NAFRI, who helped us with the field survey.

References

Andriesse E, Phommalath A (2012) Provincial poverty dynamics in Lao PDR: a case study of Savannakhet. J Curr Southeast Asian Aff 31(3):3–27. https://doi.org/10.1177/186810341203100301

Baird IG, Shoemaker B (2007) Unsettling experiences: internal resettlement and international aid agencies in Laos. Dev Chang 38:865–888. https://doi.org/10.1111/j.1467-7660.2007.00437.x

Bazzi S, Gaduh A, Rothenberg AD, Wong M (2016) Skill transferability, migration, and development: evidence from population resettlement in Indonesia. Am Econ Rev 106:2658–2698. https://doi.org/10.1257/aer.20141781

Bird K (2009) 'Voluntary' migration in Lao People's Democratic Republic. In: Overseas development institute policy brief no 3

Cole R, Rigg J (2019) Lao peasants on the move: pathways of agrarian change in Laos. Aust J Anthropol 30:160–180. https://doi.org/10.1111/taja.12312

Dommaraju P, Tan J (2014) Households in contemporary Southeast Asia. J Comp Fam Stud 45:559–580. http://www.jstor.org/stable/24339641

Ducourtieux O, Laffort JR, Sacklokham S (2005) Land policy and farming practices in Laos. Dev Chang 36:499–526. https://doi.org/10.1111/j.0012-155X.2005.00421.x

Évrard O, Baird IG (2017) The political ecology of upland/lowland relationships in Laos since 1975. In: Bouté V, Pholsena V (eds) Changing lives in Laos: Society, politics, and culture in a post-socialist state. NUS Press, Singapore, pp 165–191

Hickey M, Narendra P, Rainwater K (2013) A review of internal and regional migration policy in Southeast Asia. In: Working Papers id:5592, eSocial Sciences

High H (2008) The implications of aspirations: reconsidering resettlement in Laos. CritAl Asian Stud 40:531–550. https://doi.org/10.1080/14672710802505257

Insisienmay S, Philavanh S (2018) Internal migration in Lao PDR. In: CMER working paper series, No.0001

Manivong V, Cramb R, Newby J (2014) Rice and remittances: crop intensification versus labour migration in Southern Laos. Hum Ecol 42:367–379. https://doi.org/10.1007/s10745-014-9656-6

Nakata T (2004) Water buffalo sacrifice in a multi-ethnic village in Southern Laos. Tōnan Ajia Kenkyū (Jpn J Southeast Asian Stud) 42:74–103 (in Japanese). https://doi.org/10.20495/tak.42.1_74

Nakatsuji S (2013) Differences in the importance of shifting cultivation among villages in Northern Laos: a case study of 14 villages in Xieng Ngeun District, Luang Phabang Province. Jinbun Chiri (Jpn J Hum Geogr) 65:339–356 (in Japanese). https://doi.org/10.4200/jjhg.65.4_339

Nishimoto F (2011) Laos shōsū minzoku no kako 50 nen no jinkō fukugen (Population reconstitution in the past 50 years of ethnic minorities in Laos). Jinkō Gaku Kenkyū (J Popul Stud) 47:126 (in Japanese). https://doi.org/10.24454/jps.47.0_126_1

Niwa T, Nishimoto F, Takahashi S, Yokoyama S (2023) Rural-rural migration in a rural village, Luang prabang province, northern Laos: focusing both on immigration and emigration in the village of HB. E-J GEO 18:279–290 (in Japanese). https://doi.org/10.4157/ejgeo.18.279

Niwa T, Nishimoto F (2025) Rural–urban migration in Northern Laos: a case study of the village HB in Luang Phabang Province. In: Yokoyama S, Niwa T, Jiang H (eds) Population dynamics and livelihood changes of small-scale societies in Laos. Springer, Singapore, pp. 171–189. https://doi.org/10.1007/978-981-96-2956-5_8

Niwa T, Takahashi S, Nishimoto F (2025) Transnational labor migration from rural village in Laos to Bangkok, Thailand: focusing on relationship with the home village. In: Yokoyama S, Niwa T, Jiang H (eds) Population dynamics and livelihood changes of small-scale societies in Laos. Springer, Singapore, pp. 73–97. https://doi.org/10.1007/978-981-96-2956-5_4

Phouxay K, Tollefsen A (2011) Rural–urban migration, economic transition, and status of female industrial workers in Lao PDR. Popul Space Place 17:421–434. https://doi.org/10.1002/psp.620

Phouxay K (2010) Patterns of migration and socio-economic change in Lao PDR. Ph.D. thesis, Umeå universitet

Rigg J (2007) Moving lives: migration and livelihoods in the Lao PDR. Popul Space Place 13:163–178. https://doi.org/10.1002/psp.438

Rigg J (2019) More than rural: textures of Thailand's agrarian transformation. University of Hawaii Press, Honolulu

Rigg J, Wittayapak C (2009) Spatial integration and human transformations in the Greater Mekong subregion. In: Huang Y, Bocchi AM (eds) Reshaping economic geography in East Asia. World Bank, Washington DC, pp 79–99

Skeldon R (2006) Recent trends in migration in East and Southeast Asia. Asian Pac Migr J 15:277–293. https://doi.org/10.1177/011719680601500206

Takahashi S, Yokoyama S, Nishimoto F, Niwa T (2025) Population growth and generational response to livelihoods in a rain-fed paddy farming village of central Laos. In Yokoyama S, Niwa T, Jiang H (eds) Population dynamics and livelihood changes of small-scale societies in Laos. Springer, Singapore, pp 45–69. https://doi.org/10.1007/978-981-96-2956-5_3

Thomson SS, Baden S (1993) Women and development in Laos. Institute of Development Studies, University of Sussex, Brigton

Yokoyama S, Takahashi S, Niwa T, Nishimoto F (2025) Demographic dynamics and paddy field ownership in a remote rural area in Northern Laos. In: Yokoyama S, Niwa T, Jiang H (eds) Population dynamics and livelihood changes of small-scale societies in Laos. Springer, Singapore, pp 117–137. https://doi.org/10.1007/978-981-96-2956-5_6

Rural–Urban Migration in Northern Laos

A Case Study of HB village in Luang Phabang Province

Takahito Niwa and Futoshi Nishimoto

Abstract Previous studies on internal migration in Laos have neither considered the relationship between rural-rural migration and rural–urban migration nor sufficiently examined the processes involved in permanent migration to cities. Accordingly, this study explored the characteristics of rural–urban migration in Laos by connecting rural population dynamics with migrants' life courses. Interviews were conducted with 42 households and 22 migrants who migrated from a lowland village on the northern fringe of Laos to Vientiane. The analysis results revealed that rural-rural migration from the highlands to lowlands has been promoted as a national policy in Laos, which has significantly impacted rural population dynamics. Consequently, there appears to be active migration to cities. Those migrating to cities have entered the urban labor market in Vientiane as much-needed low-skilled workers. However, after settling in cities, they have improved their living and labor conditions by changing their jobs. Although these migrants have attempted to maintain their relationships with their villages of origin, thereby mitigating the risks associated with migration, their ties with their villages have become weaker, eventually leading to permanent departure from the village as their working and living conditions in Vientiane enhanced. Triggers for migrants to settle in urban areas are not only related to the labor market but also closely tied to their children's education.

Keywords Population migration · Rural village · Urban labor market · Life course · Laos

This chapter was a revised version of Niwa and Nishimoto (2021).

T. Niwa (✉)
College of Human and Social Sciences, Kanazawa University, Kanazawa, Japan
e-mail: niwa-2@staff.kanazawa-u.ac.jp

F. Nishimoto
Embassy of Japan in Laos, Vientiane, Laos
e-mail: nsmtfts@gmail.com

S. Yokoyama et al. (eds.), *Population Dynamics and Livelihood Changes of Small-Scale Societies in Laos*, International Perspectives in Geography 22,
https://doi.org/10.1007/978-981-96-2956-5_9

1 Introduction

As their economies become increasingly globalized, ASEAN countries continue experiencing rapid economic growth. Laos is one such country; despite the country's small size, economic growth continues[1] in the context of increasing international trade with China and Thailand and foreign direct investment (FDI). According to the October 2020 edition of the World Bank's World Economic Outlook Database, the GDP growth rate of Laos has exceeded an annual average of 7% since 2000. Similar to other Asian countries, economic growth in Laos has been accompanied by population migration from rural areas to cities. Existing research on migration has tended to be premised on the existence of megacities, such as those in Indonesia and Thailand, thereby focusing on informal sector employment and the rise of the new middle class and generalizing these issues in terms of over-urbanized cities and FDI-driven new middle class cities (Davis and Golden 1954; Konagaya 1997, etc.). Other research on migration has focused on factors other than push–pull factors from the perspective of economic disparity between urban and rural areas (Castles et al. 2014) such that migration in the rural areas of Southeast Asia is increasingly understood in terms of surplus labor and diverse livelihood strategies (Rigg 2007).

Although there appears to be a notable relationship between economic structures and internal population migration in Laos, there are geographical and social conditions that differ from those of other regions where extensive research has been conducted and are likely to affect migration. Although some research has stated that farmers in Laos have always engaged in migration (Évrard and Baird 2017), statistical data on Laos remain limited, which means that even basic information on population migration is lacking (Rigg 2007).

Therefore, this study presents a case study of a lowland farming village located in northern Laos. Based on interview surveys conducted in the home villages of migrants and the urban areas they migrated to, the authors argue that the fluid movement of population among rural areas is a contextual factor influencing the migration of populations to cities and that the process of migration to cities is deeply connected to migrants' relationships with villages.

2 Prior Research and Research Objectives

Phouxay (2010) linked the migration of populations in Laos to socioeconomic changes. Describing the characteristics of migrations that have occurred over the course of approximately a quarter of a century, Phouxay (2010) noted that the migration of populations between rural and urban areas (i.e., "rural–urban migration") began increasing in Laos around the time of the announcement of the "New Economic

[1] Laos ranks low among the 10 ASEAN countries in terms of its area of 240,000 km^2 and population of approximately 6.5 million (as of 2015). Laos has the smallest area after Singapore, Brunei, and Cambodia and the smallest population after Brunei and Singapore.

Mechanism" in 1986, which triggered the introduction of a market economy into Laos. Since the 1990s, in tandem with the development of domestic transportation infrastructure, as measures to develop rural areas and reduce poverty began being implemented by the government and international organizations, migration among different rural areas ("rural–rural migration") increased, especially migration from the highlands to lowlands, in addition to rural–urban migration. There was a notable increase in international migration in the 2000s due to the development of the global economy. Phouxay (2010) noted that migration in rural areas expanded alongside economic development but did not focus on population changes in rural areas, including rural–rural migration.

Meanwhile, Insisienmay and Philavanh (2018) used census data to clarify national trends in rural–urban migration, noting that from 1995 to 2015, migration within Laos mostly occurred within and around the capital, Vientiane, due to its more developed transportation infrastructure. However, the study did not discuss the background of this migration. Phouxay and Tollefsen (2011) discussed urban population migration from the perspective of the labor market, thus clarifying the link between the migration of women from rural areas to Vientiane and increased direct investment in urban industries and the active recruitment of women workers from rural areas. These employment opportunities have been explained as arising from the growing numbers of large-scale garment factories in Vientiane associated with multinational companies (Nolintha and Jajri 2015), with labor markets embedded in the global economy functioning as a pull factor for rural–urban migration. However, these studies only consider the urban aspect of rural–urban migration, thereby leaving the rural aspect unclear.

Cole and Rigg (2019) examined the relationship between rural communities and migration to urban areas and found that the migration of populations to cities helped diversify the livelihoods of rural households and reduce absolute poverty in Laos. However, they did not discuss the relationship between rural–urban and rural–rural migration, despite both occurring in rural areas. Inter-rural migration currently includes migration from highland to lowland areas[2]; thus, it is highly probable that the population dynamics in these rural areas affect rural–urban migration. In addition, much of the discussion on rural–urban migration has been premised on temporary migration, such as migrant work, with a lack of discussion on the processes of permanent migration from rural to urban areas and settlement in the latter.

To resolve these issues, research must reconstruct the population dynamics observed in rural areas, clarify the impact of rural–rural migration on rural–urban migration, gather systematic data on the life courses of migrants,[3] and determine the

[2] The Lao government has adopted resettlement policies to promote migration by ethnic minorities living in highland areas to the lowlands. Studies have noted that the government has sought to ban swidden agriculture and opium cultivation practiced in the highlands and shift to modernized market-oriented systems, such as plantations. However, this has often also entailed forcing ethnic minorities living in highlands, such as the Khmu and Hmong, to integrate into the ethnic Lao way of life (Bird 2009; Baird and Shoemaker 2007).

[3] The term "life course" refers to a collection of individual life histories; an important focal point of this approach is to be migrant-centric and apply due diligence in collecting the life histories of

processes by which migration occurs. Accordingly, this study clarified the characteristics of rural–urban migration in Laos by connecting rural population dynamics with the migration processes of permanent migrants.

3 Research Methods

Based on the issues described in the previous section, this study presents a case study of HB village in Luang Phabang Province and examines the characteristics of population migration in Laos. First, the study clarifies the population structure of HB village based on a comprehensive all-household survey and population reconstruction. Second, it describes population migration from HB village to Vientiane based on migrants' life courses.

Before explaining the research method, we present a brief overview of the research project that includes this study. This study is part of the research outcomes of the JSPS Grant-in-Aid research project "The mechanism of population dynamics in small-scale communities of present Laos" (2017–2021). To introduce an ecological perspective to the elucidation of the relationships among population, health, and occupation, several researchers in this project have conducted field surveys on the topics of family planning, occupation, and food acquisition in HB village and neighboring NK village alongside the study of population migration presented herein. The field survey was conducted with support from the Lao National Agriculture and Forestry Research Institute and the Lao Tropical and Public Health Institute. Based on preliminary research, which indicated that HB village was a typical paddy farming village of the kind found in the valleys of northern Laos, we judged that the village would be a suitable case study location where comprehensively observing the activities of residents, including their history of births, deaths, and past migrations, would be possible.

The first research method employed in this study was a survey of all households in HB village to clarify the population structure. Furthermore, past population dynamics were reconstructed using Nishimoto's (2011) method, which is described as follows. "We start by identifying the current ages of the household members and tracing each member back to when they first entered the household due to events such as birth or marriage, clarifying changes in the composition of households over time. We then used the family trees reconstructed using the interviews to track and record absent members who were once a part of the household but died or moved away." (Nishimoto 2011, 126) For HB village, we created records of 463 people, tracing as far back as the initial inhabitants who founded the village in the early twentieth century. Although this research method provides valuable data on rural Laos, which lacks detailed population data, it tends to produce lower population estimates as one

migrants, thereby capturing the diverse aspects of increasingly complex urban–rural relations (Soda 2008).

goes further into the past. Therefore, this study presents data only from 1975 onward, which tends to be relatively reliable.

The second research method employed was surveys of those who had migrated away from HB village to clarify the characteristics of population migration among the villagers of HB village. These surveys were conducted as structured interviews at the migration destinations. Twenty-two survey participants who migrated from HB village to Vientiane were included in the analysis. These surveys were conducted from March 5 to 11, 2019. The survey population comprised 58 people who had migrated from HB village to Vientiane. We used the following methods to contact survey participants and request their cooperation in the survey. First, we asked the survey participants' relatives and friends in HB village for the participants' contact information. We identified 41 contacts at this stage. Next, we had the relatives in HB village contact them and ask whether they would participate in the survey. Nineteen people were not interviewed because they responded that they could not participate or did not respond to multiple phone calls.

We suspect that the employment status of migrants affected the omission from the survey of many migrants. Some potential participants found it difficult to participate in the survey because they were living in the factory where they worked, whereas others may not have wanted to answer questions directly. It is also possible that some may not have wanted to participate because they were distrustful of the unfamiliar foreigners conducting the survey. Therefore, we asked the mayor of HB village to be present when we were conducting the survey in Vientiane. Meeting the village mayor alleviated some of the villagers' concerns regarding our research.

4 Population Structure of HB Village

4.1 Overview of HB Village

HB village was abandoned during the Laotian Civil War as villagers sought refuge in the forest and rebuilt at the same location after the war (Fig. 1). However, the population composition of the residents changed significantly. Therefore, this study examines population dynamics from 1975, when the civil war ended. The village is located on a small piece of flat land in a valley surrounded by mountains to the east and west. The distance to the neighboring NK village is 2 km, with a distance of approximately 4 km to the village after that (MN village). As of March 2018, basic infrastructure did not extend to HB village, with unpaved roads connecting it to surrounding villages and no mains water or electricity.

Farmland extends around the village, and all households are engaged in farming. However, not all households in HB village have paddy fields because there is no vacant land available to distribute usage rights to households who have recently moved to the village. Therefore, households that cannot afford to purchase farmland usage rights from others use any farmland they still have where they used to live or

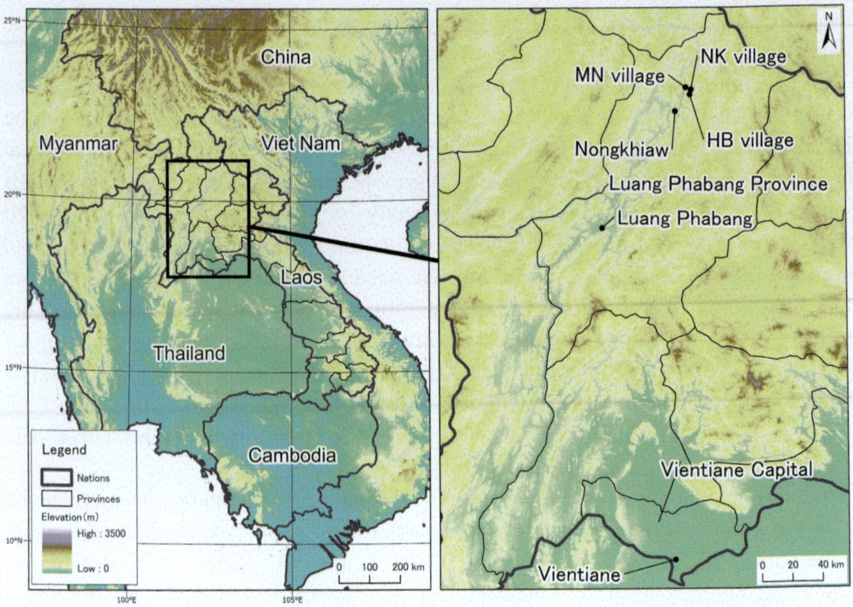

Fig. 1 Location of the field site in Luang Phabang Province, Northern Laos

practice swidden agriculture on the slopes at the foot of the mountains. Most of the harvested rice is consumed by the families who produce it. Some households keep cows and water buffaloes as a means of earning cash and as assets. One household in the village runs a guesthouse for tourists.[4]

4.2 Characteristics of the Population Structure

As of March 2018, 181 people were living in 42 households.[5] If we include 27 temporary migrants,[6] the total population is 208 (96 men and 112 women). Villagers' age composition is shown in Fig. 2, with a notably young average age of 27.3 years.[7]

Figure 3 shows changes in the village population (including temporary migrants) since 1975, organized by ethnicity. The overall trend shows a high population growth

[4] Numerous guesthouses and restaurants can be found in MN village, where many foreign tourists stay during the tourist season (the dry season).

[5] Data from the March 2018 survey were used as a starting point, with partial updates to unclear aspects of the data provided by subsequent field surveys.

[6] "Temporary migration" refers to people who leave their parents' homes to attend university or work in urban areas. Meanwhile, "permanent migration" refers to people who have migrated as a family (i.e., couples or a parent and child). These categories were created based on the terms used by the mayor of HB village.

[7] All ages presented this study are as of March 2018.

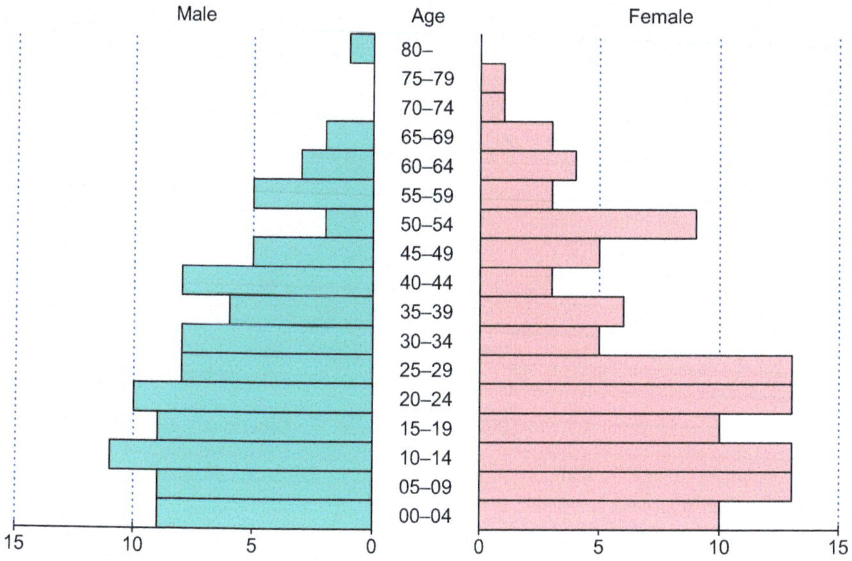

Fig. 2 Population pyramid of HB village (person). *Source* Authors' interview survey

rate until the late 1990s, with an average annual population growth rate of 5.2% until 1995. However, the population increase peaked in the late 1990s and has remained at a low rate of 1.5% since 2000. The permanent population, excluding temporary migrants, peaked at 202 in 2010 and has been declining since. In terms of ethnicity, the majority population of the village was originally ethnic Lao; their population steadily increased until the mid-1990s. The ethnic Lao population has subsequently plateaued at around 90 people. However, the Khmu population, with its origins in the highlands, has been increasing since the civil war; since the mid-2000s, the Khmu population has exceeded that of the ethnic Lao.[8]

4.3 *Changes in Population Structure*

Next, we examine changes in population structure through the lens of population dynamics. Our subjects included those who were born or died in HB village, those who moved there, and those who had lived in HB village at some point and who could be identified as having moved out by March 2018. Figure 4 summarizes these transitions in population dynamics. From 1975 to 2018, the village had 217 births, 102 deaths, 148 people moving in, and 135 people moving out. The population dynamics

[8] Children inherit their father's ethnicity. Therefore, the children of an ethnic Lao husband and Khmu wife are classified as ethnic Lao, and the children of a Khmu husband and ethnic Lao wife are classified as Khmu.

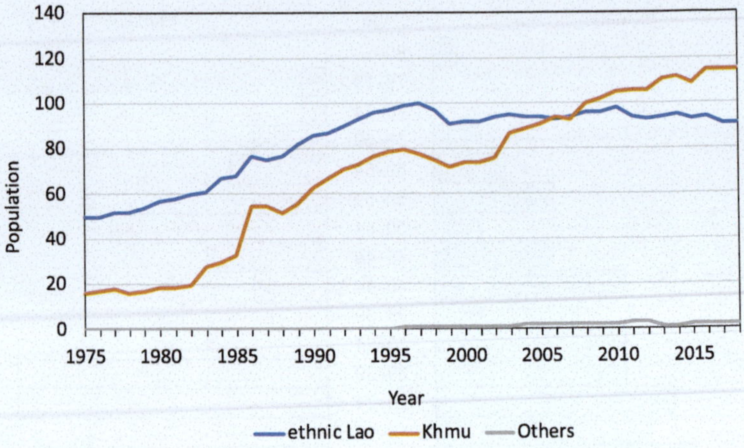

Fig. 3 Population change by ethnicity in HB village, 1975–2018. *Note* Others include Taidam, Hmong, and unknown. *Source* Authors' interview survey

changed around 2000. Until around 2000, except in 1986, natural population growth was the primary contributor to population growth, and the decrease in population in the late 1990s can be attributed to an increase in the number of deaths. In contrast, from the 2000s onwards, the increase in the number of people migrating elsewhere began exerting a stronger influence, with the population declining in the 2010s.

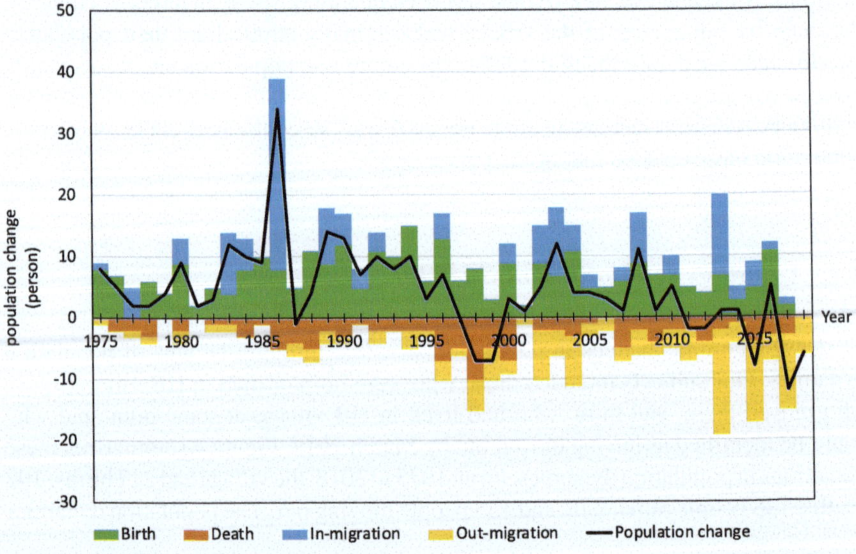

Fig. 4 Demographic trend in HB village, 1975–2018. *Source* Authors' interview survey

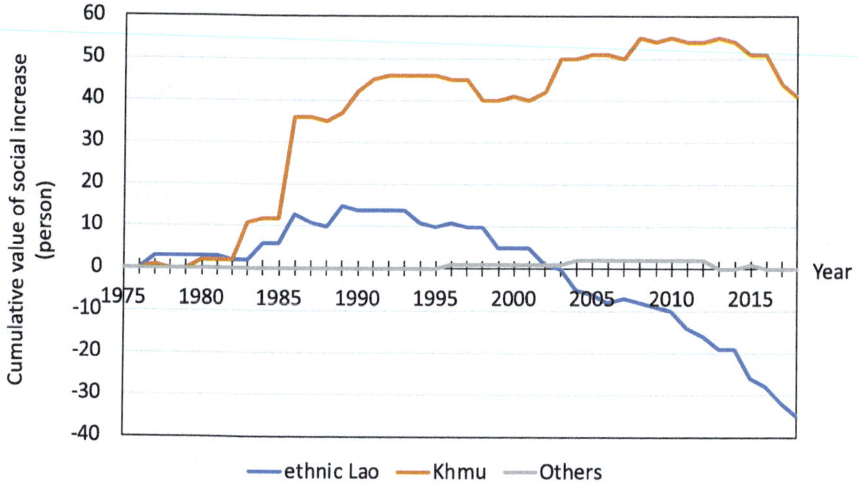

Fig. 5 Cumulative value by social increase by ethnicity in HB village, 1975–2018. *Source* Authors' interview survey

Figure 5 chronologically shows the cumulative fluctuations in ethnic group populations that make up the community, indicating that the cumulative fluctuations of the Khmu increased in the 1980s and changed little thereafter. There has been a limited influx of Khmu into HB village in recent years facilitated by the government's resettlement policy, but the number of people who have moved out has slightly exceeded this influx. Meanwhile, regarding the cumulative fluctuation in the ethnic Lao population, the state of excess immigration that existed until the early 2000s has subsequently changed to one of excess emigration. In terms of population dynamics, this can be interpreted as a basic pattern of ethnic Lao leaving and Khmu entering HB village. Moreover, regarding the chronological order of these two phenomena, the increase in the immigration of the Khmu preceded the increase in the emigration of the ethnic Lao.

5 Characteristics of the Migration of Emigrants from HB Village

5.1 Migrant Attributes

A total of 147 people emigrated from HB village, including 27 temporary migrants and 120 permanent migrants. Of the temporary migrants, 15 are Khmu and 12 are ethnic Lao; the permanent migrants comprise 32 Khmu, 85 ethnic Lao, and 3 others. This composition seems to reflect a difference in timing, wherein people leave to work away when they are young and then switch to permanent migration after establishing a basis for their livelihood at the migration destination. The most common migration destination was Vientiane, with 58 people having migrated there. In addition, 35 people migrated to urban areas within Luang Phabang Province. This makes a total of 93 people if we consider these collectively as migrations toward the capital, which means that at least 63% of those moving away from the village follow rural–urban migration.[9] This study focuses on those who have migrated to the most popular migration destination, Vientiane, and analyzes the data of 22 migrants.

We start by briefly confirming the attributes of the survey participants. In terms of the birth cohorts of the 22 participants, 3 were born in the 1960s, 2 in the 1970s, 6 in the 1980s, 10 in the 1990s, and 1 in the 2000s. Thus, there was a concentration in younger age groups. There were more men (15) than women (7). Of the 22 participants, 15 were ethnic Lao and 6 were Khmu. Most participants (17) were born in HB village. Participants from other villages had moved there with their families or after getting married.

In terms of migration status, 8 participants were temporary migrants and 14 were permanent migrants. The average age of the temporary and permanent migrants was 21.6 and 38.0 years, respectively. The difference between the two ethnic groups was significant, with the 6 Khmu comprising a majority of the temporary migrants and 13 ethnic Lao comprising a majority of the permanent migrants. Thus, in terms of ethnicity, the Khmu migrants tended to be younger.

5.2 *Migrants' Life Courses*

This section examines the life courses of the 22 surveyed participants in terms of the characteristics of population migration from HB village to Vientiane.

The 22 participants who migrated to Vientiane worked in the following industries: 7 in agriculture; 4 in manufacturing; 5 in the service industry; 3 in retail/wholesale; 2 in food services; and 1 unemployed (housewife) (Table 1). If we group industries such as services and retail into tertiary industries, this represents a significant employment opportunity for the participants, with 10 of them (almost half) employed in this sector. In addition, some of those currently working in tertiary industries previously worked in manufacturing. Regarding the relationship between the industries that participants were employed in immediately after migrating to Vientiane and those where they were currently working (Table 2), we see that 11 participants, a significant percentage, were employed in secondary industries in their first job and later switched to working in tertiary industries. In addition, although there was no clear difference among the industries in which temporary migrants worked, 8 out of 14 permanent migrants in Vientiane were working in tertiary industries. Although Vientiane borders Thailand's Nong Khai Province across the Mekong River, none of the 22 migrants had traveled to Thailand.

[9] For those who migrated to places other than Luang Phabang Province, it was not possible to determine whether they migrated to a town or village.

Table 1 Working industries of migrant workers in Vientiane Capital

Industries	Sex		Total
	Female	Male	
Primary industry (agriculture)	–	7	7
Secondary industry (manufacturing)	2	2	4
Tertiary industry	4	6	10
Service	1	4	5
Wholesale, retail	1	2	3
Restaurant	2	–	2
Non-working	1	–	1
Total	8	15	22

Source Authors' interview survey

Table 2 Changing of working industries of migrant workers in Vientiane capital

Industries working as first	Industries working as contemporary				Total
	Primary industry	Secondary industry	Tertiary industry	Non-working (housewife)	
Primary industry	3	–	1	–	4
Secondary industry	2	4	5	–	11
Tertiary industry	2	–	3	–	5
Non-working (student)	–	–	1	1	2
Total	7	4	10	1	22

Source Authors' interview survey

Here, we consider the case of Mrs. Yai (a pseudonym), a migrant worker who is a typical example of temporary migration. Mrs. Yai is of Khmu ethnicity and was born in HB village in 1998. She started working as a migrant worker at a textile factory in Vientiane in 2017 after completing four years of secondary education. Information regarding the factory was obtained from a resident of HB village who had previously worked in the area. Mrs. Yai left her job after six months[10] and temporarily returned to the village in 2018. She married another HB village native that year and returned to Vientiane as a migrant worker with her husband. She currently works with her husband at a factory manufacturing furniture parts. She lives in an employee dormitory attached to the factory and has no personal property or possessions other than a mobile phone. She and her husband send money to her parents in HB village; the money is currently being used to build a bathroom at her parents' house. Mrs. Yai evaluated her life in Vientiane as "I can earn my living, and I get new experiences

[10] Mrs. Yai stated that she left the textile factory "because it was a tough job." Another woman who is currently working at the textile factory described the harsh working environment there, "During busy periods, I work 24 h a day, broken up by only a few breaks".

Table 3 Relationships between migrant workers and HB village by migration status

Migration status	Amount of monthly income (US $)	Amount of monthly remittance (US $)	Average of monthly contacts (times)
Temporary migrants (*n* = 8)	136	26	9.0
Permanent migrants (*n* = 14)	447	14	6.6
Total	334	18	7.4

Source Authors' interview survey

at work." Thus, migrating to the city is a form of self-actualization and a part of the family's livelihood strategy.

Next, we will examine the migrants' income level. There were large income variations, with the average individual monthly income of the 22 participants being 2.88 million kip (approximately US$334)[11] and the median income being 1.5 million kip (approximately US$174) (Table 3). Income variation was also related to the degree of settlement: the income level of temporary migrants was low (at the minimum wage level in Laos), whereas permanent migrants earned on average nearly three times as much as temporary migrants. When participants shifted from temporary to permanent migration, their economic relationship with the village also changed (Table 3). More temporary migrants sent money back to the village, and they sent larger amounts than permanent migrants. Given the income differences between the two groups, temporary migrants sent 20% of their income back to the village, whereas permanent migrants sent approximately 5%.[12] Nonetheless, temporary and permanent migrants frequently (at least six times a month) made phone calls to relatives and villagers in HB village. In addition, there were cases where participants no longer had relatives in HB village but maintained contact with the village. Consequently, an ongoing relationship with the village was a characteristic feature identified in this study. Other participants had farmland in HB village.

We now discuss the case of Mr. Thongchai (a pseudonym), a permanent migrant settled in Vientiane. Mr. Thongchai, an ethnic Lao born in 1973, left HB village in 1987 for higher education in Luang Phabang Province where he trained as a teacher. After graduating, he worked as a teacher in MN village near HB village for 10 years, but in 2000, he decided to move to Vientiane because he was not earning enough as a teacher. Along with his wife's family, he moved to Vientiane where he runs a private taxi business and his wife runs a retail business. They own a house in addition to a considerable amount of personal property, such as a TV, refrigerator, and car (for

[11] Calculated at a rate of 8,614 kip to US$1. This roughly corresponds to the three-month average from January to March 2019 (https://www.xe.com/; last accessed on January 31, 2023). In some cases, participants provided only their annual household income. Here we simply calculated the respondent's income by dividing the annual household income by the number of employed people in the household.

[12] In addition to bank-facilitated remittances, participants used a remittance agency service operating in MN village or sent money with villagers who were returning to the village.

work). Mr. Thongchai assesses his current life as "comfortable" and is ensuring that his children pursue higher education. Furthermore, he has taken various measures to establish a solid basis for his livelihood, including purchasing a vacant lot next to his home in 2018 as a future home for his children. Meanwhile, his parents own some farmland in HB village, but he intends to pass this on to his siblings instead of inheriting it.

Finally, we solicited opinions from the 22 participants regarding their lives in Vientiane. These included opinions expressing satisfaction with city life, such as "it is easy to find a job," "the quality of life is good," and "it is good for the children's education."

6 Discussion

Here, we discuss the characteristics of rural–urban population migration in terms of the population dynamics in HB village and the life courses of migrants moving to Vientiane.

First, there was a large-scale influx of Khmu migrants from the highlands to HB village in the 1980s, with subsequent ethnic Lao emigration from the 2000s onward. The population structure of the village has been affected by contrasting fluctuations in the populations of the ethnic groups that constitute the community. The influx of the Khmu can be attributed to the resettlement policy of the Lao government, whereas the emigration of the ethnic Lao can be attributed to overpopulation in the village. As mentioned above, newcomers to HB village are unable to gain a share of the farmland usage rights. In a village that became overpopulated due to Khmu immigration, we can assume that the ethnic Lao quickly started moving away for work due to concerns regarding the instability of working in agriculture and settling in urban areas. Thus, rural–urban population migration in northern Laos exists against a backdrop of rural–rural migration triggered by the government resettlement policy.

Second, Vientiane offers employment opportunities in the manufacturing industry in unskilled positions that do not require prior experience and in agriculture, which facilitates a work style similar to participants' livelihoods in HB village.[13] Thus, migration from rural to urban areas is possible. After moving to the city, many participants changed jobs in search of a better working environment and life and came to work in the service industry. Changing jobs in this manner has been made possible by employment expansion in the service industry within the increasingly urbanized Vientiane labor market. Among the rural–urban migrants in our study, few worked in the informal sector and none of them had work styles similar to the new middle class, unlike in prior studies that focused on over-urbanized cities. We found that rural–urban migration from northern fringes of the nation to the capital is

[13] Some people farm their own farmland, some rent farmland, and others are employed at large-scale plantations. Land prices have risen in recent years, making it difficult to purchase farmland when migrating elsewhere.

supported by the formal sector of the service industry, which has expanded in tandem with urban growth in Vientiane.

Third, Vientiane is regarded as the final destination of rural–urban migration. Migrants from HB village do not tend to consider employment in Thailand as an option after arriving in Vientiane. We can infer this from the fact that many of those who wish to settle down have already acquired real estate in Vientiane and established a foundation for their lives there. The migrants also express a high degree of satisfaction with urban life, which can be considered an important factor in the switch from temporary to permanent migration. This satisfaction also appears to be associated with their children's attainment of higher education, including the second generation, which contributes to their settlement in cities. Nonetheless, the migrants maintain their relationships with HB village; some still have real estate there. Rather than attributing this to a wish to leave a legacy there for their descendants, maintaining a relationship with the village appears to stem from the desire to retain the option of returning to the village should there be any problems with work or life in Vientiane. This can be regarded as a safety net for first-generation immigrants. However, this thinking seems to weaken as the migrants establish a stronger foundation for life in Vientiane. The fact that the migrants maintain their relationship with the village is a similar feature to the case of paddy field farming villages in central Laos (Niwa et al. 2025). On the other hand, it is significantly different from the case in central Laos in that Vientiane, rather than Bangkok, Thailand, is the final destination. This difference may be influenced by the villages' difficulty in obtaining information about Thailand and the lack of a human network of chain migration in the fringe farming villages of northern Laos.

7 Conclusion

This study examined the characteristics of rural–urban population migration in Laos based on a case study of a lowland farming village located on the northern fringes of Laos by connecting population dynamics within the village with the migration processes of migrants.

The results of the analysis can be summarized as follows. Laos has a national policy of promoting rural–rural migration from highland to lowland areas. This policy has had a particular impact on the northern region, which comprises both areas. As a result, it is causing overpopulation in rural areas. This overpopulation has been associated with increasing instability in agriculture, which causes rural–urban migration with a steady influx into Vientiane by those who are able to move to the city. Migrants enter the labor market in Vientiane as low-skilled workers. However, as they settle in cities, they improve their living and labor conditions by changing jobs. However, they seek to maintain relationships with their village of origin to mitigate any risks associated with migration. As they develop stronger life foundations in Vientiane, their ties with the village become weaker and they become permanent migrants. Triggers for migrants to settle in urban areas are related to the labor market and their children's education.

Finally, adopting a macroscopic perspective of the national structure of Laos to examine the characteristics of rural–urban population migration, which has previously been examined from a microscopic perspective, we note that without improving the living conditions in rural areas, the migration of populations from rural to urban areas will continue unabated. The major factors that underlie rural–urban migration include the persistent instability of agriculture as means of livelihood in rural areas, lackluster results of government efforts to revitalize rural areas, and rapid economic growth of urban areas. Further research is required on the characteristics of rural–urban migration and the changes that have occurred in urban and rural areas due to migration.

Acknowledgements This study was supported from JSPS Grants-in-Aid JP17H01633. We would like to thank everyone involved who cooperated with the field survey. We would also appreciatethank. the assistance of Dr. Khamla Sengphaxaiyalarh in NAFRI, who helped us with the field survey.

References

Bird K (2009) 'Voluntary' migration in Lao people's democratic republic. Overseas Dev Inst Policy Br 3:1–9

Baird IG, Shoemaker B (2007) Unsettling experiences: internal resettlement and international aid agencies in Laos. Dev Chang 38:865–888. https://doi.org/10.1111/j.1467-7660.2007.00437.x

Castles S, de Haas H, Miller MJ (2014) The age of migration: international population movements in the modern world, 5th edn. The Guilford Press, New York

Cole R, Rigg J (2019) Lao peasants on the move: pathways of agrarian change in Laos. Aust J Anthropol 30:160–180. https://doi.org/10.1111/taja.12312

Davis K, Golden HH (1954) Urbanization and the development of pre-industrial areas. Econ Dev Cult Change 3:6–26. https://www.jstor.org/stable/1151656

Évrard O, Baird IG (2017) The political ecology of upland/lowland relationships in Laos since 1975. In: Bouté V, Pholsena V (eds) Changing lives in Laos: society, politics, and culture in a post-socialist state. NUS Press, Singapore, pp 165–191

Insisienmay S, Philavanh S (2018) Internal migration in Lao PDR. CMER Working Paper Series 1, pp 1–34

Konagaya K (1997) Urban economy and urban structure of Asian mega-cities. Kikan Keizai Kenkyū (Quart J Econ Stud) 20:61–89 (in Japanese)

Nishimoto F (2011) Laos shōsū minzoku no kako 50 nen no jinkō fukugen (Population reconstitution in the past 50 years of ethnic minorities in Laos). Jinkō Gaku Kenkyū (J Popul Stud) 47:126 (in Japanese). https://doi.org/10.24454/jps.47.0_126_1

Niwa T, Nishimoto F (2021) Internal migration in rural Laos: a case study of the village H in Luang Prabang Province. Jinkō Gaku Kenkyū (J Popul Stud) 44:21–32 (in Japanese). https://doi.org/10.24454/jps.2102001

Niwa T, Takahashi S, Nishimoto F (2025) Transnational labor migration from rural village in Laos to Bangkok, Thailand. In: Yokoyama S, Niwa T, Jiang H (eds) Booktitle. Springer, Place, pp 99–116. https://doi.org/10.1007/978-981-96-2956-5_5

Nolintha V, Jajri I (2015) The garment industry in Laos: technological capabilities, global production chains and competitiveness. ERIA Discuss Paper Ser 2015–13:1–32

Phouxay K (2010) Patterns of migration and socio-economic change in Lao PDR. Ph.D thesis, Umeå University

Phouxay K, Tollefsen A (2011) Rural–urban migration, economic transition, and status of female industrial workers in Lao PDR. Popul Space Place 17:421–434. https://doi.org/10.1002/psp.620

Rigg J (2007) Moving lives: migration and livelihoods in the Lao PDR. Popul Space Place 13:163–178. https://doi.org/10.1002/psp.438

Soda R (2008) Approaches to rethinking rural-urban migration in Southeast Asia: a case of the Iban in Sarawak, Malaysia. E-J GEO 3(1):1–17 (in Japanese). https://doi.org/10.4157/ejgeo.3.1_1

Aspects of Modern Contraceptive Use Among Women in Rural Northern Laos

Chihiro Shirakawa and Yoshie Moriki

Abstract The purpose of this chapter is to clarify the aspects of modern contraceptive use among women in rural northern Laos and to examine the characteristics thereof. To that end, the authors conducted questionnaire-based interviews with all 180 adult female residents (at the time of the survey) of NK village, Luang Phabang Province. The results indicated that almost half the adult women in NK village have experience with modern contraceptive methods; in particular, well over half of the younger women (in their 20s and 30s) were found to have experience using these methods. The main method in use was oral contraception (the pill). Knowledge about this and other modern contraceptive methods was mainly obtained from medical personnel and many of the women had begun using this contraception based on recommendations from medical personnel. Additionally, while the middle-aged women (in their 40s and 50s), who had numerous children already, used modern contraceptive methods in order to avoid further pregnancies, the younger women used them as planned birth control from a relatively young age, indicating a difference in usage format between the generations.

Keywords Modern contraceptive use · Women · Lao · Khmu · Laos

1 Introduction

July 11 is World Population Day. The World Population Prospects 2022, issued on this day, estimate that the world's population will reach 8 billion on November 15, 2022. It is anticipated that the global population will reach 8.5 billion by 2030 and

C. Shirakawa (✉)
Graduate School of Human Sciences, Osaka University, Suita, Japan
e-mail: itakuma@hus.osaka-u.ac.jp

Y. Moriki
College of Liberal Arts, International Christian University, Mitaka, Japan
e-mail: moriki@icu.ac.jp

S. Yokoyama et al. (eds.), *Population Dynamics and Livelihood Changes of Small-Scale Societies in Laos*, International Perspectives in Geography 22,
https://doi.org/10.1007/978-981-96-2956-5_10

9.7 billion by 2050.[1] While the global population is increasing; however, in some countries and regions, birth rates are falling and the rate of population growth is slowing or shifting towards decreasing trends. Laos is one of the countries with a notably decreasing birth rate.

One of the factors cited for this decrease is the ease of obtaining modern contraceptive methods, which are directly related to childbirth (Aoyama et al. 2001: 127). With this in mind, it is essential to understand the trends concerning the use of modern contraceptive methods when exploring the background of the decreasing Laotian birth rate. In the case of Laos, information can be obtained from the results of governmental surveys and those done by international organizations (e.g. Ministry of Health and Lao Statistics Bureau 2012: 85–94; UNFPA 2015; Lao Statistics Bureau 2018: 90–98). However, these provide only large-scale perspectives on population trends at the macro level of countries, which makes it difficult to obtain a detailed understanding of individuals' specific usage of modern contraceptive methods. To do that, research focusing on micro units, such as specific communities or small groups, is required. However, little research of this kind has been done heretofore in Laos (e.g. Sychareun et al. 2013; Shimazawa 2014; Chantakoumane et al. 2020).

In order to fill in these blanks, this chapter clarifies the aspects of the usage of modern contraceptive methods by women in a northern Laos farming village (below, NK village) and analyzes their characteristics. Note that "modern contraceptive methods" here refers specifically to the use of oral contraceptives (the pill), the injection of contraceptive hormones, the use of intrauterine contraceptive devices (the coil), tubal ligation surgery, and other similar methods. The chapter focuses on women because women in Laos have been found to use modern contraceptive methods at a notably higher rate than men (Shimazawa 2014: 214; Chantakoumane et al. 2020: 77; see also Shirakawa et al. 2025). In addition, the data presented in this chapter is derived, as discussed below in Sect. 3, from questionnaire-based interviews conducted with all the adult women in NK village. Research based on an exhaustive survey of a specific community is extremely rare in Laos, suggesting that this chapter has a trailblazing significance.

The following section presents an overview of NK village in Sect. 2, an explanation of the survey methods in Sect. 3, and a detailed discussion of the attributes of the survey targets in Sect. 4. Thereafter, the survey targets are divided into two categories: those with and without experience using modern contraceptive methods. The latter category is discussed in Sect. 5, while the former is discussed in Sect. 6. Finally, based on the aforementioned results, the various observed characteristics of modern contraceptive use by the women of NK village are discussed, and concluded in the Sect. 7.

[1] https://www.unic.or.jp/news_press/info/44737/, accessed November 20, 2022.

2 Study Area

The NK village, the study area for this chapter, is located in the northern region of Luang Phabang Province in northern Laos. It is roughly four kilometers from MN village, which faces the Ou River as it flows through the province from north to south and plays host to numerous foreign tourists. The road between the two villages is unpaved; the villages are without public electricity, with most households using micro-hydrogenerators installed on the nearby river.

As of 2018, the population of NK village was 354 people, including 169 men and 185 women. With regard to ethnic groups, the Lao constitutes the majority, with 229 people (approximately 64.7% of the total village population), while the Khmu accounts for 125 people (approximately 35.3%). According to the Lao villagers, NK village, which is believed to date back almost 470 years, was originally inhabited by the Lao people. In 1974, however, the Khmu people moved there from mountain villages some three hours away on foot, making it a multiethnic settlement.

Almost all the residents are engaged in agricultural activities, such as paddy and swidden farming. The total area of paddy in the NK village is 45.5 hectares. It should be noted, however, that not all households hold rice paddies. Those who do not, or who hold only small areas of paddy, derive their livelihood through swidden agriculture and gathering forest products, such as cardamom and galangal. Additionally, many households engage in livestock raising, with the village maintaining a total of 270 cows, 120 water buffaloes, and 50 pigs.[2]

MN village has several shops offering a comparatively wide range of foodstuffs and everyday commodities. Additionally, a periodic market is held every ten days. In contrast, there are no large shops in NK village. While Vietnamese traders do traverse the area, the inaccessibility to NK village during the rainy season results in a decline in their activity. Therefore, villagers rely on the shops in MN village for their everyday needs. Travel between NK village and MN village is mainly on foot or by motorcycle; all the households in the village own a motorcycle. However, there are only two cars in the whole village.

There are no modern medical facilities in NK village. The nearest medical facility is the health center in MN village. This health center is regularly staffed by multiple medical personnel, including nurses and a doctor. Alongside seeing and treating outpatients, these medical personnel make monthly visits to the 12 villages under the health center's jurisdiction, including NK village, for gynecological care. In addition, vaccines are administrated to children under the age of five every three months in these villages.

In terms of institutional educational facilities, the village has only an elementary school. Therefore, children who progress to junior high school are required to attend the junior high school in MN village.

---

[2] The numbers of animals and previously indicated areas of rice paddies are as of 2017.

3 Research Methods

This chapter represents part of the research outcomes of the JSPS Grant-in-Aid research project "The mechanism of population dynamics in small-scale communities of present Laos."[3] This project is intended to clarify the mechanisms of population dynamics in small-scale communities, with emphasis on the relationships between various aspects, such as migrations, livelihoods, reproduction, and medical care. To that end, researchers from various fields, including demography, human ecology, geography, and cultural anthropology, collaborated with the Lao Tropical and Public Health Institute and the National Agriculture and Forestry Research Institute to conduct multiple surveys of NK village and the neighboring HB village.

The data presented in this chapter was obtained from a two-day survey conducted in March 2019 in NK village by the authors, Chihiro Shirakawa and Yoshie Moriki. Prior to this survey, the authors conducted a preliminary survey in NK village in August 2017. The 2019 survey targeted all the adult female residents of the village at the time for questionnaire-based interviews. The questionnaire was composed of three sections: general questions about target attributes, questions about usage of modern contraceptive methods, and questions about marital relations. The total number of questions was approximately 30. For the interviews, the authors used printouts of the translation of the English questionnaire into Lao.

The interviews were conducted by six Laotians including a female staff member from the Lao Tropical and Public Health Institute, a male doctor and nurse from the MN village health center, a male nurse from the district hospital, a female traditional midwife from NK village, and a male university professor from Luang Phabang. The six interviewers conducted individual interviews with the women gathered in the village square, filling out the Lao questionnaires with the information they received. The six interviewers included people with whom the women were familiar, such as personnel from the MN village health center and the NK village traditional midwife. Additionally, as most of interviewers were young and treated the women with warmth, the interviews were conducted in an extremely calm and pleasant atmosphere. The authors provided assistance with the interview process when necessary.

4 Survey Targets

The survey was conducted on all adult women (aged 18 and older) residing in NK village at the time. This resulted in data from 99 subjects. Of these, however, 19 residents were absent from the village during the survey, and their answers were

[3] Research on this project spanned a period from AY2017 through AY2022, led by the editor of this collection, Satoshi Yokoyama. Originally, the project was to run through AY2020, but as travel to Laos became difficult due to the global spread of the novel coronavirus, the research period was extended for 2 years through AY2022.

Table 1 Number of subjects by age group

Age group	Number of people	Ratio (%)
20s	12	15.0
30s	24	30.0
40s	15	18.8
50s	15	18.8
60s	9	11.3
70s	4	5.0
80s	1	1.0
Total	80	99.9

Source Authors' questionnaire-based interviews

provided by their mothers or others close associates, resulting in insufficient interview responses. Therefore, this chapter examines interview response data for 80 subjects.

Table 1 shows the age of the 80 subjects at the time of the survey by age range. As the table indicates, the largest number of subjects were in their 30s, followed by those in their 40s and 50s. These three age ranges accounted for a majority of the survey targets.

In terms of ethnicity, the subjects included 51 Lao (approximately 63.8% of the total 80 subjects)[4] and 25 Khmu (approximately 31.3%) women. As noted in Sect. 2, the percentage of Lao and Khmu within the village as a whole is approximately 64.7% to 35.3%. Consequently, the ratio of Lao to Khmu among the survey targets reflected that of the village as a whole.

Regarding occupational status, 66 of the subjects were engaged in agricultural activities as their primary occupation, dominating the study population. The remaining subjects were distributed as follows: five were housewives, one engaged in a small business, and eight responded with "other." Because the content of "other" is unclear, the number of subjects not engaged in agriculture may in fact be even higher. Regardless, it is clear that the majority of subjects were engaged in agricultural activities.

The remaining attributes are marital status, number of children, and number of births experienced. Of the total study population, 77 were married, accounting for a majority of the participants (approximately 96.3%). Of these, 68 had spouses at the time of the survey, while the remaining nine were either divorced or widowed. Only two women were unmarried.[5]

The number of children the subjects had at the time of the survey is shown in Table 2. The most common response was two children, followed by three and four children. Incidentally, the average number of children per subject was approximately 3.1. However, the number of children as shown in Table 2 also includes adopted children; it does not include children who died or were given away for adoption after birth.

[4] There were also 4 subjects who did not respond or whose responses were unclear.

[5] There was also 1 subject who did not respond.

Table 2 Number of subjects based on number of children

Number of children	Number of people
0	4
1	10
2	25
3	11
4	11
5	8
6	3
7	4
8	1
9	1
No response	2
Total	80

Source Authors' questionnaire-based interviews

Taking these points into account, Table 3 shows the number of births experienced by the subjects at the time of the survey. As the table shows, the most common number was three, followed by two and four. The average number of childbirths per subject was 4.4. Among the subjects, 51 had given birth to more children than they were raising at the time. This is thought to be due to the common occurrence of stillbirths, children dying after birth, and children given away for adoption.

Finally, the subjects' educational background must be taken into account. As shown in Table 4, the largest number of subjects were elementary school graduates, followed by those who had attended elementary school but had not graduated, that is, those who did not finish elementary school. The next largest number was those with no schooling at all. Additionally, just ten subjects had had received education beyond the elementary level. A total of 69 subjects, roughly 86.3% of the whole, had not attended school at all, had not finished elementary school, or had completed their education with an elementary school graduation. This means that, if these three groups are considered to have minimal education, a large majority of the subjects fall into this category.

Based on Table 4, the subjects in this category of minimal education were mostly in their 40s and older. Among the women who were middle-aged and elderly, just one had received an education beyond the elementary level. In contrast, while five of the women in their 20s had minimal education, six had received education beyond elementary level. This suggests that younger women in the subjects were inclined to have relatively higher levels of education.

Table 3 Number of births among the subjects

Number of births	Number of people
0	2
1	6
2	10
3	19
4	10
5	8
6	6
7	5
8	5
9	2
10	3
11	1
12	1
No response	2
Total	80

Source Authors' questionnaire-based interviews

Table 4 Educational background and age group of subjects

Educational background	Number of people	20s	30s	40s	50s	60s	70s	80s
None	18	1	4	5	2	2	4	0
Did not finish elementary school	21	0	7	3	5	5	0	1
Elementary school graduate	30	4	11	7	6	2	0	0
Did not finish junior high school	6	3	1	0	1	0	0	0
Junior high school graduate	3	2	1	0	0	0	0	0
High school or above	1	1	0	0	0	0	0	0
No response	1	1	0	0	0	0	0	0
Total	80	12	24	15	14	10	4	1

Source Authors' questionnaire-based interviews

5 Subjects with no Experience of the Use of Modern Contraceptive Methods

This section and the following one indicate and discuss data on aspects of modern contraceptive use among the women constituting the survey targets. Of the 80 subjects, 37(approximately 46.3% of the whole) had experience with the use of modern contraceptive methods. Of these, 25 used them on a regular basis, ten had used them in the past, and the remaining two used them from time to time. Forty subjects (half of the whole) had never used modern contraception and constituted a majority of the group.[6] This section details information on participants with no experience of modern contraception.

The subjects were first analyzed by age range (Table 5), and it was found that the 40 subjects in this group were most commonly in their 40s, followed by those in their 50s and 60s. When analyzed by percentage, over 60% of those in their 40s and older had never used modern contraception. This group included all the subjects in their 60s and 80s.[7] That is, over half the middle-aged and elderly subjects had no experience with modern contraception, and in particular, almost none of participants in their 60s and older had ever used it. The results of the statistical analysis confirmed the correlation between age and experience. Chi-square and Fisher's exact tests found that the middle-aged and elderly women over 40 were significantly more likely to have no experience of modern contraception, with a significance level of 1% ($\chi^2 = 28.56$, $p < 0.001$).[8] However, we must note that many of the younger women, those in their 20s and 30s, also lacked this experience.

Moving on, the 40 women in this group included 22 Lao, 17 Khmu, and one non-respondent (Table 5). As noted in the previous section, the total number of subjects was 51 Lao and 25 Khmu. Therefore, the ratio of those without experience was approximately 43.1% for Lao and 68% for Khmu, indicating a notably high rate of those without experience among the Khmu. Statistical analysis confirmed the correlation between ethnicity and experience. Chi-square and Fisher's exact tests found that Khmu were significantly more likely than Lao to be without experience of modern contraception, with a significance level of 5% ($\chi^2 = 4.68$, $p = 0.030$).

However, it is not appropriate to jump to the conclusion that Khmu women tend not to use modern contraceptive methods. This is because the Lao and Khmu subject groups did not cover the same proportion of age ranges.

As shown in Table 6, the Lao subjects were most commonly in their 30s, followed by those in their 40s. These 2 age groups constituted more than half of the 51 Lao subjects. In contrast, the Khmu subjects were most commonly in their 50s and 60s, with these 2 age groups constituting almost half of the 25 Khmu subjects. In addition, the percentage of those in their 70s and 80s was also higher in the Khmu group than

[6] There were also 3 subjects who did not respond.

[7] Regarding the 70s age group, 3 of the 4 subjects had no experience. The experience of the remaining subject was unclear.

[8] Cooperation for the statistical analysis concerning this chapter was provided by Satoshi Yokoyama. We would like to express our deep gratitude.

Table 5 Number of subjects without experience of modern contraceptive use by age group and ethnicity

Age group	Number of survey targets	Number of subjects without experience	Percentage (%) of subjects without experience	Number of Lao subjects without experience	Number of Khmu subjects without experience
20s	12	2	16.7	2	0
30s	24	5	20.8	3	2
40s	15	11	73.3	8	3
50s	15	9	60.0	5	4
60s	9	9	100.0	3	6
70s	4	3	75.0	1	1
80s	1	1	100.0	0	1
Total	80	40	50.0	22	17

Source Authors' questionnaire-based interviews

in the Lao group, with, a very low percentage of those in their 20s and 30s. In short, in comparison with the Lao group, the Khmu group was characterized by a high percentage of middle-aged and elderly women. In addition, as noted above, over half the middle-aged and elderly subjects had no experience with modern contraceptive methods. Notably, almost none of those in their 60s and older had ever used them. Based on these points, the large numbers of those without experience among the Khmu group can be understood as a reflection of the age group composition thereof, that is the large numbers of middle-aged and elderly women.

Next, let us consider educational background. As shown in Table 7, of the 40 women in this group, the largest number were elementary school graduates, followed

Table 6 Age groups within Lao and Khmu ethnicities

Age group	Lao (number of subjects)	Percentage of each age group within Lao subjects overall (%)	Khmu (number of subjects)	Percentage of each age group within Khmu subjects overall (%)
20s	8	15.7	2	8
30s	20	39.2	3	12
40s	10	19.6	5	20
50s	9	17.6	6	24
60s	3	5.9	6	24
70s	1	2.0	2	8
80s	0	0.0	1	4
Total	51	100.0	25	100

Source Authors' questionnaire-based interviews

Table 7 Educational background and age groups of subjects with no experience of the use of modern contraceptive methods

Educational background	Number of people	20s	30s	40s	50s	60s	70s	80s
None	12	0	1	4	2	2	3	0
Did not finish elementary school	8	0	0	1	1	5	0	1
Elementary school graduate	17	1	3	6	5	2	0	0
Did not finish junior high school	2	1	0	0	1	0	0	0
Junior high school graduate	1	0	1	0	0	0	0	0
Total	40	2	5	11	9	9	3	1

Source Authors' questionnaire-based interviews

by those with no schooling at all. The previous section combined the subjects who had not attended school at all, those who had not finished elementary school, and those whose schooling had ended with elementary school graduation into the minimal education group. This resulted in a total of 37 women who did not have experience with modern contraceptives and had only minimal education. This point suggests that most of those without experience of modern contraceptive methods are those with minimal education. However, as the table shows, those without experience also include subjects with relatively high educational levels, i.e. having attended or graduated from junior high school. It is important to note that the lack of experience of modern contraception is not necessarily limited to those with minimal education.

The survey also included questions regarding the reasons behind the lack of experience with modern contraceptive methods. This section concludes with a consideration of these responses. Of the 40 subjects in this group, responses could be obtained from 36 subjects; of these, 29 responded along the lines of "leaving it to nature," constituting a majority. Among these respondents, the largest number (8) were in their 60s, followed by seven each in their 40s and 50s. While middle-aged and late-middle-aged women were numerous here, there were also four women in their 30s and one in her 20s, indicating a small number of relatively younger women who had the same reason for not using modern contraceptive methods.[9]

In consideration of the respondents' educational background, the largest number in this group (13) were elementary school graduates, followed by nine with no schooling at all and five who had not finished elementary school. Therefore, those with minimal education formed a majority. However, the group also included one who had not finished junior high school and one who had graduated, indicating women with relatively higher education levels had also responded that they were "leaving it to nature." In addition, the next common reason was "wanting children," with five respondents.

[9] The remaining two subjects were in their 70s.

Table 8 Number of subjects with experience of the use of modern contraceptive methods by age group and ethnicity

Age group	Number of survey targets	Number of subjects with experience	Percentage (%) of subjects with experience	Number of Lao subjects with experience	Number of Khmu subjects with experience
20s	12	8	66.7	5	2
30s	24	19	79.1	17	1
40s	15	4	26.7	2	2
50s	15	6	40.0	4	2
60s	9	0	0.0	0	0
70s	4	0	0.0	0	0
80s	1	0	0.0	0	0
Total	80	37	46.3	28	7

Source Authors' questionnaire-based interviews

6 Subjects with Experience of the Use of Modern Contraceptive Methods

6.1 Attributes of Subjects with Experience

In this section, we will examine subjects with experience using modern contraceptive methods. As noted at the beginning of the previous section, 37 out of the 80 subjects had this experience. Of these, 25 used them on a regular basis, ten had used them in the past, and the remaining two used them from time to time.

First, let us analyze the ages of these 37 subjects. As shown in Table 8, the most common age range was the 30s, followed by the 20s. The percentage of those with experience was notably high in both groups. Additionally, a considerable number of subjects in their 40s and 50s reported experience, with a notable 40% of those in their 50s. On the other hand, no respondents in their 60s or older had experience with modern contraception. As noted in the previous section, statistical analysis confirmed the correlation between age and experience. Chi-square and Fisher's exact tests found that younger women in their 20s and 30s were significantly more likely to have experience. Incidentally, all six women in their 50s as well as two women in their 40s and two in their 30s responded that while they had experience using modern contraception in the past, they were not using it at the time of the survey.

In terms of ethnicity, these 37 subjects, as shown in Table 8, 28 were Lao and seven Khmu.[10] As 51 of the total of 80 subjects were Lao and 25 Khmu, approximately 54.9% among the Lao and 28% of the Khmu had experience, indicating a lower rate of experience among the Khmu. As noted in the previous section, statistical analysis confirmed the correlation between ethnicity and experience. Chi-square and Fisher's

[10] There were also two subjects whose response was unclear.

Table 9 Educational background and age groups of subjects with experience of the use of modern contraceptive methods

Educational background	Number of people	20s	30s	40s	50s
None	5	1	3	1	0
Did not finish elementary school	13	0	7	2	4
Elementary school graduate	13	3	8	1	1
Did not finish junior high school	4	2	1	0	1
Junior high school graduate	2	2	0	0	0
Total	37	8	19	4	6

Source Authors' questionnaire-based interviews

exact tests found that Lao subjects were significantly more likely than Khmu to have experience.

However, as pointed out in the previous section, the Khmu subjects were characterized by a higher percentage of middle-aged and elderly women compared to the Lao. As shown in Table 6, there were nine Khmu women aged 60 and older with no experience at all. Conversely, there were only 5 among the younger women in their 20s and 30s with experience (Table 6). Based on these points, the lower number of subjects with experience among the Khmu compared to the Lao can be attributed to the subjects' age rather than their ethnicity, as middle-aged and elderly subjects with no experience were more common among the Khmu.

We will now turn our attention to educational background. As shown in Table 9, the largest group among the 37 subjects were those who had attended elementary school and elementary school graduates. Section 4 combined these 2 groups with the subjects who had not attended school at all into the minimal education group, resulting here in a total of 31, a large majority of the 37 subjects. There were also as many as six subjects with relatively high educational backgrounds (attending or graduating from junior high school) who had experience using modern contraceptive methods. These results suggest that these methods are in use among the subjects regardless of educational background. In addition, statistical analysis using Chi-square and Fisher's exact tests confirmed that there was no significant difference or correlation between educational background and experience.

6.2 Methods Used and Context of Their Use

This subsection addresses, in order, the methods used, ways of acquiring knowledge thereof, and reasons for starting use for the 37 subjects with experience of the use of modern contraceptive methods.

Regarding the methods used by the 37 subjects, the most commonly used method was the pill (17 subjects), followed by both the pill and contraceptive hormone injections (three subjects), and injections alone (one subject). When viewed by type, 20 used the pill and four used injections, suggesting that the use of the pill was common among the subjects. Additionally, none of them had used intrauterine contraception (the coil) or undergone tubal ligation surgery.

Because the survey did not ask specific questions about how the pill and other methods were acquired, that information could not be obtained. According to the doctor at the MN village health center, however, the health center began providing pills, condoms, and injections free of charge upon request in October 2018, the year before the survey. Before that point, payment was required. The price of a month for the pill or an injection with a three-month efficacy period were both 5,000 kip.[11] The injections must be given by a doctor, and cannot be used easily at any time by the women involved. Therefore, the pill was apparently more popular due to its ease of use. The provision of pills and provision of injections also took place when the doctor visited NK village for infant vaccinations, pregnancy checks, and so on. Keeping this in mind, it can be assumed that the NK village residents using pills and injections, as noted above, may include those obtaining them via the MN village health center.

The question thus arises as to how women acquire knowledge of the modern contraceptive methods mentioned above? The most common response to the question on this topic, supplied by 18 subjects, was through the health center staff. The "health center" here is thought to refer to the MN village health center. Next most common was hospital staff, with four responses. This hospital is thought to be the district hospital in Nong Khiaw, a town a little over an hour down the Ou River from MN village. Additionally, three subjects responded with parents and three with friends.

The results indicate that the MN village health center, which is relatively close to NK village, holds a particularly important position as a way for women to acquire knowledge of modern contraceptive methods. It is also notable that most of the subjects obtained knowledge from medical personnel affiliated with medical facilities, such as the health center or the hospital. Incidentally, when seeing outpatients or performing vaccinations and gynecological care in the villages, the medical personnel from the MN village health center also conduct health education on family planning, including the use of modern contraceptive methods.

The role of medical personnel at the health center and other institutions cannot be ignored, in terms of provision of information as well as opportunities to begin use of the pill and other contraceptives. In response to the survey's question on reasons for using modern contraceptive methods, as many as 12 subjects responded that medical personnel had recommended their use. Other responses included how easy contraception was to obtain (four subjects), a recommendation from a friend (one subject), for better health (one subject), and other content (two subjects). These results indicate that medical personnel at the health center and elsewhere are not only the main sources of knowledge, but provide major triggers for the start of usage of modern contraceptive methods.

[11] The exchange rate of the local currency, the kip, to the Japanese yen was 1.5 yen to 100 kip as of April 2017.

6.3 Timing of Beginning Usage

This subsection focuses on the timing of when the women began using modern contraceptive methods, in particular with relation to age. Of the 37 women who had experience of modern contraceptive methods, ten were relatively older women, with four in their 40s and six in their 50s. Of these, six were able to provide information on the timing when they began using contraception, of whom again 4 (one in her 40s, three in their 50s) began using it after giving birth to three or more children. In contrast, there were two (one each in their 40s and 50s) who began using contraception after marriage before their first child or after the birth of their second child.

Of the 19 subjects in their 30s with experience, 11 were able to ascertain information on timing. Of these 11, the largest group was the five who had begun use after their first child, followed by three who had begun use after their second child. This age group also included one subject who had begun using modern contraception before marriage. In contrast, there were only two subjects who had begun use after giving birth to three or more children. Incidentally, both of these were close to age 40, at 38 and 39, respectively.

In the remaining eight subjects in their 20s with experience, six were able to ascertain information on timing. Of these, the largest group was the three subjects who had begun use after marriage before giving birth to their first child. The other three had begun before marriage, after their first child was born, and after their second child was born.

The aforementioned information indicates that among middle-aged women, modern contraceptive methods have mainly been used by multiparas with numerous children as a method of preventing the conception of any more children. In contrast, among younger women, these methods have mainly been used from a relatively young age as a means of planned birth control.

7 Discussion and Conclusion

In light of the discussion so far, this section lists out the observed characteristics of the aspects of modern contraceptive use by the women of NK village, discusses these findings to some extent, and concludes the chapter. First, the following 6 characteristics were observed:

(1) Almost half the adult women of NK village have experience of the use of modern contraceptive methods.

(2) This experience is especially common among younger women in their 20s and 30s; well over half the women in this age range have experience of using modern contraception. Additionally, while not as commonly, a certain number of the women in their 40s and 50s also had this experience. In contrast, none of the women in their 60s or older had this experience.

(3) Those with experience included both Lao and Khmu women, as well as those with relatively high educational backgrounds and those with no schooling at all. It is difficult to suggest that this experience is more or less likely based on ethnicity or educational background.
(4) The main contraceptive used was the pill. Injections were also used to some extent, but no evidence of the use of the coil or tubal ligation surgery was found.
(5) The women mainly acquired knowledge of modern contraceptive methods from medical personnel at the health center and other such institutions. Accordingly, many of the women had begun using contraception based on recommendations from medical personnel. This indicates that medical personnel play a major role in the adoption of modern contraceptive methods among the women of NK village.
(6) Among middle-aged women in their 40s and 50s, modern contraceptive methods were largely used by women with multiple children in order to prevent further pregnancies. In contrast, among younger women in their 20s and 30s, these methods were mainly used from a relatively young on as a means of planned birth control, indicating a difference in usage format between age groups.

Among these 6 characteristics, characteristics (5) and (6) also apply to KH village in central Laos, as discussed in Shirakawa et al. (2025). Among the women of KH village, modern contraceptive methods first came into use among multiparas with multiple children in order to avoid further pregnancies. Thereafter, they spread to younger women as a method of planned birth control from an earlier age immediately after marriage. Medical personnel played a major role in both acquisition of knowledge on and usage of modern contraceptive methods for the women of KH village.

Conversely, the NK village survey was unable to obtain information on the specific number of years of use of modern contraceptive methods. Therefore, this survey cannot provide a chronology of the adoption of modern contraceptive methods, as was done in the KH village survey in Shirakawa et al. (2025). However, with the case of KH village in mind, it may be possible to postulate that in NK village, as discussed in this chapter as well, multiparas with multiple children were the first to acquire knowledge and recommendations from medical personnel and thus begin the use of modern contraceptive methods. This subsequently led to younger women also began to adopt contraceptives as a method of planned birth control.

In any case, the use of modern contraceptive methods in NK village has become well accepted, centering on younger women in their 20s and 30s. What, then, are the social factors affecting the use of these methods by so many women there? The topographical feature of NK village results in alimited availability of land suitable for new paddy reclamation in the basin. Nevertheless, the population has increased due to the influx of Khmu people and other factors, resulting in a significant number of residents who do not own rice paddies (Yokoyama et al. 2023: 305–306). For these people in particular, life in NK village is economically difficult, more so for those with more children. This may play a part in the acceptance of the use of modern contraceptive methods. However, no empirical information on this factor was obtained in the authors' survey, therefore, it remains a hypothesis and this examination is left to future study.

It is noteworthy that in KH village in central Laos, discussed in Shirakawa et al. (2025), overseas work in Thailand is common, which has a definite influence on the acceptance process of modern contraceptive methods in the village. On the other hand, overseas labor is almost unheard of in NK village. However, as detailed in Yokoyama et al. (2023: 294–295), migrating for work or permanent residence to domestic urban areas such as Luang Phabang or Vientiane has become more common from the late 1990s on. That said, information clarifying the influence of this shift was not obtained from the authors' survey in NK village. Consequently, it remains unclear whether there is a relationship between migration for work or residence from NK village to Laotian cities and the trends of modern contraceptive use by the women of NK village. If this relationship is irrelevant, it is necessary to identify the reasons why. These questions likewise must be left to future research.

Acknowledgements This study was supported from JSPS Grants-in-Aid JP17H01633. In writing this chapter, we received tremendous support from many people and related organizations. First, we would like to extend heartfelt thanks to the people of NK village for their patient participation in the survey that formed the basis of this chapter. Wide-ranging cooperation was provided during the survey by Ms. Hongkham Xayyavong, Mr. Phetsakhone Phetinsy, Mr. Phoummy Phongsavat, Mr. Anoukone Bouphasyli, and Mr. Somthapheuy Syvongsak. The support of the Lao Tropical and Public Health Institute and the National Agriculture and Forestry Research Institute was indispensable in conducting this survey. We would like to express our deep gratitude to everyone involved.

References

Aoyama A, Hara H, Kita E (2001) Health, development, and gender: well-being for all. Yūhikaku, Tokyo (in Japanese)

Chantakoumane G, Maguet C, Essink D (2020) Married couples' dynamics, gender attitudes and contraception use in Savannakhet Province, Lao PDR. Glob Health Action 13:74–84. https://doi.org/10.1080/16549716.2020.1777713

Lao Statistics Bureau (2018) Lao social indicator survey II 2017. Vientiane

Ministry of Health & Lao Statistics Bureau (2012) Lao social indicator survey 2011–12. Vientiane

Shimazawa K (2014) Raosu ni okeru "seishoku kontorōru" no yōsō: Josei no kenkō purojekuto toshite no dōnyū (Aspects of "reproductive control" in Lao PDR: its introduction as a health project for women). In Kohama M, Matsuoka E (eds) Childbirth and family planning in Asia: the politics of women's reproductive bodies. Benseishuppan, Tokyo, pp 192–222 (in Japanese)

Shirakawa C, Nishimoto F, Takahashi S, Yokoyama S (2025) Growth in the use of modern contraceptive methods and the related social context in a central Laotian farming village. In: Yokoyama S, Niwa T, Jiang H (eds) Population dynamics and livelihood changes of small-scale societies in Laos. Springer, Singapore, pp 99–116. https://doi.org/10.1007/978-981-96-2956-5_6

Sychareun V, Hansana V, Phengsavanh A, Phongsavan K (2013) Awareness and attitudes towards emergency contraceptive pills among young people in the entertainment places, Vientiane City, Lao PDR. BMC Women's Health 13:14–23. https://doi.org/10.1186/1472-6874-13-14

United Nations Population Fund (UNFPA) (2015) Family planning situation analysis: Lao people's democratic republic. UNFPA Lao PDR, Vientiane

Yokoyama S, Takahashi S, Niwa T, Nishimoto F (2023) Population dynamics and paddy holdings in remote rural villages of northern Laos. E J GEO 18:291–308 (in Japanese). https://doi.org/10.4157/ejgeo.18.291

Family Planning Methods and Fertility Level in a Northern Remote Village of Laos: From Viewpoints of Women's Autonomy and International Health Cooperation

Yoshie Moriki and Chihiro Shirakawa

Abstract This chapter sheds light on the future of contraceptive use and fertility reduction in the context of international medical cooperation in Laos. The authors conducted a survey on differences in fertility behavior and contraceptive use by generations among married women aged 18 years and older in NK village, MN district, northern Laos in 2019. The analysis revealed a clear generational decline in fertility in NK village. First, young adults in their 20 s and 30 s had statistically significantly lower numbers of births and surviving children than older adults in their 40 s and older. It was also found that the proportion of contraceptive usage-experience was statistically significantly higher among the younger generation than that of the older one. The type of contraceptive used was either oral contraception or injectable contraceptive hormones, regardless of the generation. The fertility level in NK village is already at a level where further decline should be feared. Thus, the impacts of the government's commitment to expanding the use of long-acting reversible types of contraceptives need to be carefully evaluated should the replacement level of fertility be maintained. In addition, from the perspective of reproductive health and the rights of women, it is better to promote those methods of contraception that allow women to be more in control of their bodies rather than types of contraception that require more intensive medical intervention.

Keywords Fertility · Family planning methods · Laos · Reproductive health/rights · International health cooperation

---

Y. Moriki (✉)
College of Liberal Arts, International Christian University, Mitaka, Japan
e-mail: moriki@icu.ac.jp

C. Shirakawa
Graduate School of Human Sciences, Osaka University, Suita, Japan
e-mail: itakuma@hus.osaka-u.ac.jp

S. Yokoyama et al. (eds.), *Population Dynamics and Livelihood Changes of Small-Scale Societies in Laos*, International Perspectives in Geography 22,
https://doi.org/10.1007/978-981-96-2956-5_11

1 Introduction

In view of demographic dynamics of the Lao People's Democratic Republic ("Laos") and movements in international medical cooperation, this chapter examines contraceptive use and fertility reduction in NK village of MN district in northern Lao. Laos is in the mountainous region of Southeast Asia and remains one of the least developed countries (Ministry of Foreign Affairs, Japan 2022). Historically, it is a place that has undergone drastic social and economic changes, such as civil wars during the Cold War era, the abolition of the monarchy and the establishment of a republic, the transition to a socialist system, and the subsequent development of economic policies for opening up to international markets (Yamada 2018). In terms of demographics, the total population has increased from around 5.2 million in 2000 to around 7.2 million in 2021, while the total fertility rate has decreased from 5.3 in 1995 to 4.5 in 2005, 3.2 in 2015, and 2.5 in 2020 (United Nations 2002, 2008, 2017, 2021). The infant mortality rate has more than halved from 90 in 1995 to 42 in 2015, and down to 35 in 2020 (United Nations Inter-agency Group for Child Mortality Estimation 2023). Although the maternal mortality rate, a significant indicator for the state of women's reproductive health, has decreased significantly from 544 in 2000 to 292 in 2010 and 185 in 2017 (World Health Organization 2019), there is still room for improvement. These indicators suggest that the population of Laos is currently in the final phase of the last stage of the demographic transition, i.e., although the fertility rate and mortality rate are on the decline, neither has reached a low level.

The introduction and promotion of family planning in Laos have been closely linked to the movement of international aid. According to Shimazawa, family planning as a population policy began in the 1990s when Western aid was offered to Laos. The notable point was that it changed the concept of family planning from "controlling the number of children" to "adjusting birth spacing" as it is difficult for the Lao government to accept and promote the intention of population control held by aiding countries; the country has a small population size and ensuring a steady population growth was a crucial task (Shimazawa 2007). Since then, the Lao government has continued to work with donors such as the United Nations Population Fund (UNFPA) and the Japan International Cooperation Agency (JICA) to promote family planning as a national policy. One example is JICA's health promotion program for mothers and children, which set up a system to distribute contraceptives such as condoms and oral contraceptives while also providing vaccinations and vitamins to infants (Department of Human Development, the Japan International Cooperation Agency 2006).

Furthermore, family planning in Laos has been influenced by the global interest in reproductive health and rights prompted by the 1994 International Conference on Population and Development, known as the Cairo Conference. The concept of reproductive health and rights, which advocates the right to enjoy sexual and reproductive health, has been a major paradigm shift in population policy as it encouraged the autonomous decision-making of individuals, especially women. Based on this respect for self-determination, women have been encouraged to choose whether or not to have children, and (if so) when, at what intervals, and how many children to have (Sato and Ikegami 2012). Following these global trends, a number of policies and laws were implemented throughout the 2000s, and now the National Reproductive Health Maternal and Child Health (RMNCH) Integrated Strategy and Action Plan 2016–2025 has been introduced. This action plan sets detailed numerical targets, including contraceptive coverage, delivery rates in healthcare facilities, and maternal health checkup rates. (Ministry of Health, Lao People's Democratic Republic 2016).

Essentially, what the concept of reproductive health and rights advocates is that the women themselves can control their own sexual and reproductive health under their autonomy. It is argued, however, that in the reality of Laos, fertility behavior based on women's autonomy is not always practiced. According to Shimazawa (2014), who has long been involved in the field of reproductive medicine in Laos, the use of condoms among married couples is low due to the social perception that condoms are used when having sex with "other women" outside of marriage in a male-dominated gender relationship. She also reports that the use of dangerous back-street abortions and Chinese-made oral abortion pills outside of a doctors' supervision is rampant as abortion is illegal. In addition, Iinuma (2013) points out that there are organizational limitations from a gender perspective with regard to the Lao Women's Union, which has played a major role in guiding the distribution of contraceptives and birth spacing in Laos. In socialist Laos, women's participation in society has been encouraged to help build a state as a labor force, while traditional gender roles and gender consciousness have been maintained. In other words, maternal protection through the adjustment of birth spacing through family planning is first and foremost important for its significance in protecting the health of women as workers who contribute to the state.

Population policy is a visible form of state control over the population. The respect for individual agency in sexuality and reproduction proposed at the Cairo Conference was therefore groundbreaking. Matsuoka (2017) argues that the scope of individual autonomy accepted by society is defined by various means, not only through the number of children one is expected to have but also through such issues as recommendations for specific places to give birth. In such a social context, the Lao government agreed to and signed an international initiative called the Global Family Planning 2020 (FP2020) in 2016. The Lao government has pledged to increase the contraceptive prevalence rate from 42 to 65% by 2020, to reduce the unmet needs for modern contraceptive methods to 13%, and to increase the prevalence of long-term reversible contraceptive methods such as intrauterine devices (IUDs) and implants (Family Planning 2030 2016). The FP2020 stemmed from the vision proposed at the London Family Planning Summit in 2012 hosted by the UK Government, the Bill & Miranda Gates Foundation, the United Nations Population Fund (UNFPA), and the governments that receive aid. The initiative aimed to provide access to contraceptives, information, and services to 120 million women and girls living in the least developed countries by 2020. Its successor is called the FP2030, which is currently being implemented with a focus on data collection related to the achievement levels of targets (Family Planning 2030 2021). The notable aspect of the initiative is that it promotes long-acting reversible contraception. The Laos branch of UNFPA explains that one of the reasons for the slow increase in contraceptive use in Laos is the use of short-acting contraceptives, such as oral contraceptives and contraceptive hormone injections (UNFPA 2019). However, whether the promotion of thorough contraceptive methods by long-acting forms is appropriate as a population policy in the current Lao demographic conditions is open to examination.

Okuno (2006) argues that the introduction of systemized modern medicine into the non-Western world was one of the aspects of colonialism's governing techniques and discusses the structure of its domination from the perspective of imperial healthcare. In the post-World War II world, he points out that imperial medicine was transformed into the development of "charity medicine" led by American foundations and argues that charity medicine activities became the foundation for later international medical cooperation. Keeping the above issues in mind, this chapter questions the future of family planning implementation in Laos, based on survey conducted by the authors.

2 Methods

The authors conducted a door-to-door survey in March 2019 in NK village in northern Laos in collaboration with the Lao Tropical and Public Health Institute and the National Agriculture and Forestry Research Institute. The survey was part of a research project that aimed to examine the population dynamics of small-scale social groups in Laos. The research project had obtained ethics approval from the Ministry of Health National Ethics Committee for Health Research (ID 2018.22.MP), and this survey portion of the research had been additionally approved by the ethical review committee at the institution to which the lead author belongs. The Lao interviewers explained the protection of personal information and the voluntary nature of survey participation to the survey participants in advance, both orally and in writing, using a consent form translated into Lao.

A total of 8 people conducted the survey: 1 officer of the Lao Tropical and Public Health Institute sent from Vientiane, 5 healthcare members, including a doctor from the MG village health center, and 2 Japanese researchers (the authors). The survey was designed by the authors in English and translated into Lao. The survey team traveled on a modified tractor for about 30 min on an unpaved road to NK village to collect data from the villagers. On the first day, the team members explained the purpose of the survey to the village leaders and re-obtained their consent to allow us to enter the village to conduct the survey. After that, we decided to use the meeting place of the Lao Women's Union (LWU) located in the middle of the village as the interview base. For data collection, the team also asked a village midwife and members of LWU to reach out to village women to encourage them to participate in the survey.

The study's target group was married women who were 18 years old or older and residing in NK village. The village was in the midst of a construction boom at the time, especially with the influx of Khmu families that had emigrated from the mountain areas. As a result, it was difficult to calculate the exact number of married women in the village. The family records of the health center had a total of 102 women aged 18 or above (both married and unmarried) registered as of January 2019. The authors attempted to contact all these women as much as possible using those family records during the data collection period of 2 days we spent in the village. As a result, the team managed to interview 99 women, including 19 cases wherein their husbands, mothers, or other family members answered the survey on behalf of the women who were away from the village due to work or studies. The authors excluded those who were unmarried or non-residents (including ones answered by family members) and one case whose interview was incomplete/insufficient and used all the remaining 77 cases for the analysis. A statistical software SPSS was used for the analyses included in this chapter. Therefore, it can be argued that the survey covered almost all the women in the village.

The interview was conducted face-to-face with the interviewees. As the interviews involved private aspects of women's lives, the interviewers made sure to be considerate of the interviewees' responses. Nonetheless, there was no uncomfortable atmosphere that the researchers were worried would arise. The interviews proceeded in a relaxed mood, and sometimes, even laughter was heard. At the end of the interviews, we took a picture of each of the interviewees with an instant camera and handed them the photos as a gesture to thank them for participating in the survey. This way of expressing gratitude was more appreciated than expected, and not only young but also senior women seemed to enjoy having their photos taken.

3 Results

Table 1 summarizes the basic attributes of married women in NK village. Their average age was 44, and most (88%) of them were with their spouses (12% were separated or widowed). Ethnically, 64% were ethnic Lao, 33% were Khmu, and 3% were others (including the unknown), which shows that different ethnic groups co-inhabit the village. Most of their education was up to the elementary school level (27% dropped out and 38% graduated), while 23% had no education and only 12% progressed to junior high school (8% dropped out and 4% graduated). Regarding occupations, most of them were farmers (84%). Since the other answers (housewives: 7%, small business owners: 1%, and others: 8%) formed a small proportion, it is apparent that NK is a village based on agriculture. As reported by Niwa and Nishimoto (2021), who conducted a survey of H village, a village adjacent to NK village, these villages are surrounded by mountains, and villagers carry out farming around the hamlets located on a few flat lands. The roads leading to nearby MN village are unpaved, and there is no public infrastructure for water or electricity.

Next, Table 2 shows the distributions of women by age categories. The breakdown was as follows: those in their 20 s and 30 s (women in prime reproductive ages) comprised 43% and the rest (57%) were in their 40 s and 50 s or older (Table 2). The average number of births among different generations was 3.1 for the younger generation (in their 20 s and 30 s) and 5.6 for the older generation (in their 40 s

Table 1 Basic attributes of survey participants

Attributes		Number of people	%
Marital status	Married	68	88.3
	Separated	9	11.7
Ethnicity	Lao	49	63.6
	Khmu	25	32.5
	Others	2	2.6
	Unknown	1	1.3
Education	None	18	23.4
	Did not finish elementary school	21	27.3
	Elementary school graduate	29	37.7
	Did not finish junior high school	6	7.8
	Junior high school graduate	3	3.9
Occupation	Farmer	5	84.4
	Housewife	65	6.5
	Small business	1	1.3
	Others	6	7.8

Source Authors' questionnaire-based interviews

or over). The number of births in the former was statistically significantly lower (average difference: 2.490, $t = 4.904$, $p < 0.001$). Similarly, the number of surviving children was significantly lower for the younger generation (2.4) than that for the older generation (3.6) (average difference: 1.273, $t = 3.222$, $p < 0.002$) (Table 3). It should be noted that women in the younger generation may not have completed their reproductive life, although the usage of contraceptives discussed below indicates that their fertility intention has been achieved.

Tables 4 and 5 show the differences in contraceptive use experiences between generations. First of all, the proportion of those who answered that they always use contraceptive measures was overwhelmingly high among those in their 20 s (70%). Those who use contraceptive measures irregularly or not at all were a minority (10% and 20%, respectively). Similarly, 65% of those in their 30 s always use contraceptive

Table 2 Distributions of women by age categories

Age group	Number of people	%
20s	10	13.0
30s	23	29.9
40s	15	19.5
50s +	29	37.7
Total	77	100.0

Source Authors' questionnaire-based interviews

Table 3 Number of children by generation

	20–30s ($n = 33$)	40s + ($n = 44$)	Difference in average	Degree of freedom	t	p
Number of births	102	240	2.5	63.443	4.904	<0.001
Mean	3.1	5.6				
SD	1.4	2.9				
Surviving number of children(1)	78	160	1.3	65.699	3.222	<0.002
Mean	2.4	3.7				
SD	1.1	2.3				

Source Authors' questionnaire-based interviews
Note (1) One respondent who was 40 years of age or older did not respond to the question about the number of surviving children

measures (4% use them irregularly), 9% have used them but no longer do, and 22% do not. Regarding those in their 40 s and 50 s, those who always use them comprise 13% and 0%, respectively, and those who have never used them at all comprise 73% and 76%. Regarding the older generations, 13% in their 40 s and 21% in their 50 s have stopped using them, which suggests that some have stopped using them due to changes in their reproductive function (Table 4). The comparison of the experiences of using contraceptive measures (always, sometimes, and no longer) between the younger and the older generation shows that the proportion of those who have used them is statistically significantly higher among the younger generation ($\chi^2 = 23.09$, $p < 0.001$) (Table5). The types of contraception used were oral contraception (pills) or contraceptive hormone injections at any stage of their lives (from pre-marriage to postmenopausal), regardless of age.

Finally, the survey also asked about the frequency of sexual intercourse with the spouse. As shown in Table 6, both young and old generations were sexually active. All of those in their 20 s had sexual intercourse once a week or more, and about 80% of those in their 30 s and 40 s maintained a similar frequency. For those in their 50 s,

Table 4 Contraceptive use by generation

	20s (%)	30s (%)	40s (%)	50s + (%)	Total (%)
Always use	7 (70.0)	15 (65.2)	2 (13.3)	0 (0.0)	24 (31.2)
Used irregularly	1 (10.0)	1 (4.3)	0 (0.0)	0 (0.0)	2 (2.6)
Used but stopped	0 (0.0)	2 (8.7)	2 (13.3)	6 (20.7)	10 (13.0)
Never used	2 (20.0)	5 (21.7)	11 (73.3)	22 (75.9)	40 (51.9)
Unknown	0 (0.0)	0 (0.0)	0 (0.0)	1 (3.4)	1 (1.3)
Total	10	23	15	29	77

Source Authors' questionnaire-based interviews

Table 5 Contraceptive use experience by generation

	20–30s (%)	40s + (%)	χ^2	P
Have used	26 (78.8)	10 (23.2)	χ^2 (1) = 23.095	p <0.001
Never used	7 (21.2)	33 (76.7)		

Source Authors' questionnaire-based interviews

Table 6 Frequency of intercourse with the spouse by generation

	20s (%)	30s (%)	40s (%)	50s + (%)	Total (%)
More than 2–3 times/week	8 (80.0)	11 (47.8)	5 (33.3)	3 (10.3)	27 (35.1)
Once a week	2 (20.0)	7 (30.4)	8 (53.3)	4 (13.8)	21 (27.3)
Once every two weeks	0 (0.0)	2 (8.7)	1 (6.7)	1 (3.4)	4 (5.2)
Once a month	0 (0.0)	1 (4.3)	1 (6.7)	11 (37.9)	13 (16.9)
Not at all	0 (0.0)	1 (4.3)	0 (0.0)	4 (13.8)	5 (6.5)
Unknown	0 (0.0)	1 (4.3)	0 (0.0)	6 (20.7)	8 (9.1)
Total	10	23	15	29	78

Source Authors' questionnaire-based interviews

although the proportion of those who said once a week or more dropped to 24%, more than half (65%) still had sexual intercourse once a month.

4 Discussion and Conclusion

This chapter sheds light on family planning measures and fertility levels in a northern village of Laos using the example of NK village. The fertility rate in Laos has been declining. The results of the NK survey confirm that fertility in the NK village has also been decreasing. The number of children born to young people in their 20 s and 30 s has declined compared to older people in their 40 s and above, both in terms of the number of births and the number of surviving children. As for the younger generation, the number of children may still increase in the future, but NK Village data show that the average number of children in their 20 s is 2.0, and 70% use contraception at all times (10% use it irregularly). Therefore, it is inferred that the fertility level of these younger women will not nearly reach that of older women.

Although the detailed process of fertility reduction in comparison with other parts of Laos is beyond the scope of this study, it should be noted that the contraceptive methods used in the village, i.e., oral contraceptives and hormonal injections, are consistent with those reported in a central farming village (Shirakawa et al. 2025). Furthermore, a clear lack of condoms as a birth control method among married couples is another similarity found. Avoidance of male condoms and heavy reliance on female-dependent methods have been reported by Chanthakoumane et al (2020), based on a study in Savannakhet Province, Laos. The study argues that the use of contraceptives is regarded as a woman's duty in male-oriented social orders. Their findings, then, further point out that wife battering is considered acceptable by many of their respondents, suggesting the presence of gender inequality in social norms. Regarding wife battering, another study by Tran et al (2016) has similarly reported a considerable level of acceptance of intimate partner violence by both men and women in Laos, based on the 4th round of the Multiple Indicator Cluster Surveys (MICS). The indicators used in the survey included several questions asking whether a husband is justified in beating his partner in various situations:(wife) going out without telling him, neglecting the children, arguing with him, refusing to have sex with him, and burning the food. Moreover, a summary report of the Lao Social Indicator Survey (LSIS) 2011–2012 shows that over half of respondents, regardless of gender, educational level, age, or place of residence, accept wife battering for any of the above reasons. As for the wife's refusal to have sex with the husband, about 25 percent agreed that a wife beating is acceptable (Ministry of Health and Lao Statistics Bureau 2012). These studies indicate the imbalance in gender relations and show that power inequality is reflected in the daily lives of women, including the sphere of sexuality and reproduction.

The NK survey results found the frequency of sexual intercourse with the spouse, a proximate determinant of fertility, is well maintained. Thus, in terms of fecundability, a high frequency of sexual intercourse between spouses is a positive element in ensuring a reasonable level of fertility. Given the above reports regarding gender inequality, however, the meaning of the high frequency of sexual intercourse may need to be critically interpreted. During the fieldwork in 2020, the first author conducted interviews with married couples in NK village and asked about different activities a couple is supposed to do together. One activity that was consistently mentioned as a "couple activity" was to sleep in the same place every night. In addition, several interviewees mentioned that when a couple is not sleeping together over three days, neighbors and other villagers start to worry about the couple's relationship. Unfortunately, regarding sexuality, there is no quantitative data except for the frequencies of sexual intercourse, and no information was obtained about qualitative aspects of sexual activities between spouses, including issues related to intimate partner violence in the village. It is important in the future to investigate the relationships between the avoidance of condom use along with the reliance on female-heavy contraceptive methods and the high frequencies of spousal sexual intercourse in the context of male-oriented social organizations.

In conclusion, the findings from the NK study show that the fertility rate is reaching a level where the authorities should be concerned about a further decline. Demographers have already observed the trends where the fertility rate declined beyond the level of population maintenance in other Asian countries, such as Japan and China (Suzuki 2012). Therefore, at this point in the reduction stage, recommending the appropriate family planning methods is important if maintaining the population size is a significant issue for the country, and expanding the use of long-acting forms of contraception should be carefully examined. Further, it is concerning from the perspective of reproductive health and rights that the government has been promoting the use of IUDs and implants, which are not easy methods for women to manage by themselves. It is more difficult for women to stop or start using them without medical assistance, especially in remote locations such as NK Village. This study has found that the fertility rate has been steadily declining due to the use of conventional short-acting forms of contraception, that is, oral contraceptives and hormone injections. Therefore, given the living environment of village women, it is better to promote those methods of contraception that allow women to be more in control of their bodies rather than moving to long-acting types that require more intensive medical intervention. Moreover, the lack of condom use should be critically questioned not only for the benefits of preventing pregnancy and sexually transmitted diseases but also for achieving fairer reproductive responsibilities between genders. As Chanthakoumane et al (2020) state, it is time to start more actively involving men in the issue of family planning toward building better gender-balanced sexual relations. Finally, although NK village is sustained by farming in a remote area, it does not mean it is unaffected by impacts from outside. In this sense, NK village has been affected by the government policy of migrating people from mountainous areas to the NK village. Differences in reproductive behaviors across ethnicities are one urgent topic that needs to be examined in future studies.

Acknowledgements We received cooperation from many people and relevant organizations for our research. First, we want to express our deep gratitude to the people of NK village who were kind enough to participate in the survey. In particular, support from a traditional midwife and women from the Lao Women's Union is greatly appreciated. During the field survey, the diligence and cheerfulness of the Laotian staff, including Ms. Hongkham Xayavong, Mr. Phetsakhone Phetinsy, Mr. Phoummy Phongsavat, Mr. Anoukone Bouphasyli, and Mr. Somthapheuy Syvongsak was a great help. We conducted the survey due to the collaboration with Lao Tropical and Public Health Institute and the National Agriculture and Forestry Research Institute. We conducted this research with a grant from the JSPS Grants-in-Aid for Scientific Research (A) (Project No. 17H01633; Principal Investigator: Satoshi Yokoyama).

References

Chanthakoumane K, Maguet C, Essink D (2020) Married couples' dynamics, gender attitudes and contraception use in Savannnakhet Province, Lao PDR. Glob Health Action 13:74–84

Department of Human Development, the Japan International Cooperation Agency (2006) Lao people's democratic republic special medical equipment provision (vaccination expansion plan/ population and family planning) equipment planning survey report. https://openjicareport.jica.go.jp/pdf/11836343_01.pdf. Accessed 23 Oct 2022. (in Japanese)

Family Planning 2030 (2016) Lao PDR FP 2020 commitment. Supporting commitment documents. https://fp2030.org/lao-peoples-democratic-republic. Accessed 23 Nov 2022

Family Planning 2030 (2021) Accountability, our history. Family planning 2030. https://fp2030.org/about?targetTab=our-history. Accessed 23 Nov 2022

Iinuma T (2013) Jendā byōdō to syakaisyugi soshiki: Kaihatsu kyōryokuka no Raosu josei dōmei no hensen wo tōshite (Gender equality and socialist organization: With respect to the transformation of the Lao Women's Union under the development cooperation regime). In: Suzuki M,

Takimura T (eds) Kokusai kaihatsu to kyōdō: NGO no yakuwari to jendā no shiten (international development and collaboration: the role of NGOs and gender perspective). Akashi shoten, Tokyo, pp 245–265. (in Japanese)
Matsuoka E (2017) Kodomo wo umu · Kazoku wo tukuru jinruigaku: Orutanatibu eno sasoi (Anthropology of giving birth and creating a family: an invitation to the alternative). Benseishuppan, Tokyo. (in Japanese)
Ministry of Foreign Affairs, Japan (2022) Bōeki to kaihatsu (Trade and development). https://www.mofa.go.jp/mofaj/gaiko/ohrlls/ldc_teigi.html. Accessed 23 Nov 2022. (in Japanese)
Ministry of Health, Lao People's Democratic Republic (2016) National strategy and action plan for integrated services on reproductive, maternal, new born and child health 2016–2025. http://www.laoshealth.org/assets/national-rmnch-strategy-2016-2025.pdf. Accessed 25 Nov 2022
Ministry of Health, Lao Statistic Bureau (2012) LAO PDR Lao Social Indicator Survey (LSIS) 2011–12 (multiple indicator cluster survey/demographic and Health Survey) https://lao.unfpa.org/sites/default/files/pub-pdf/LSIS%20Report%20English%202011-2012.pdf. Accessed 1 July 2024
Niwa T, Nishimoto F (2021) Internal migration in rural Laos: a case study of the village H in Luang Prabang Province. Jinkō Gaku Kenkyū (The Journal of Population Studies) 44:21–32 (in Japanese)
Okuno K (2006) Teikoku iryō to jinruigaku (Imuperial healthcare and anthropology). Shumpūsha Publishing, Kanagawa (in Japanese)
Sato R, Ikegami K (2012) Syussyoryoku-tenkan to ripurodakuthibu herusu/raitsu (Fertility transition and reproductive health and rights). In: Atoh M, Sato R (eds) Sekai no Jinko-kaihatsu Mondai (The issue of population development in the world). Harashobō, Tokyo, pp 137–174. (in Japanese)
Shimazawa K (2007) Ripurodakushon: "Umukoto" wa tanjun dewa nainoka? (Reproduction: Isn't it simple to give birth?) In: Ikeda M, Okuno K (eds), Iryō jinruigaku no Ressun: Yamai wo meguru bunka wo saguru (lessons in medical anthropology: exploring cultures of illness). Gakuyōshobō, Tokyo, pp 148–171. (in Japanese)
Shimazawa K (2014) Raosu ni okeru "seishoku kontorōru" no yōsō: Josei no kenkō purojekuto toshite no dōnyū (Aspects of "reproductive control" in Lao PDR: Its introduction as a health project for women). In: Kohama M, Matsuoka E (eds), Childbirth and family planning in Asia: the politics of women's reproductive bodies. Benseishuppan, Tokyo, pp 192–222. (in Japanese)
Shirakawa C, Nishimoto F, Takahashi S, Yokoyama S (2025) Growth in the use of modern contraceptive methods and the related social context in a central Laotian farming village. In: Yokoyama S, Niwa T, Jiang H (eds), Population dynamics and livelihood changes of small-scale societies in Laos. Springer, Singapore, pp. 117–137. https://doi.org/10.1007/978-981-96-2956-5_6
Suzuki T (2012) Very low fertility in Eastern Asia and Europe: trends, determinants and policy responses. Jinkō Mondai Kenkyū (j Popul Probl) 68(3):14–31
Tran TD, Nguyen H, Fisher J (2016) Attitude towards intimate partner violence against women among women and men in 39 low-and middle-income countries. PLoS ONE 11(11):e0167438. https://doi.org/10.1371/journal.pone.0167438
UNFPA (2019) Tracking progress of family planning programme. UNFPA, Lao People's Demographic Republic. https://lao.unfpa.org/en/news/tracking-progress-family-planning-programme. Accessed 23 Nov 2022
United Nations (2002) Demographic yearbook 2000. United Nations Publication, New York
United Nations (2008) Demographic yearbook 2005. United Nations Publication, New York
United Nations (2017) Demographic yearbook 2016. United Nations Publication, New York
United Nations (2021) Demographic yearbook 2020. United Nations Publication, New York
United Nations Inter-agency Group for Child Mortality Estimation (2023) Infant mortality rate-total. https://childmortality.org/all-cause-mortality/data?indicator=MRY0&refArea=LAO. Accessed 6 February 2025.

World Health Organization (2019) Trends in maternal mortality 2000 to 2017 estimates by WHO, UNICEF, UNFPA, World Bank Group and the United Nations population division: executive summary. https://apps.who.int/iris/bitstream/handle/10665/327596/WHO-RHR-19.23-eng.pdf?sequence=13&isAllowed=y. Accessed 25 Nov 2022
Yamada N (2018) Raosu no kisochisiki (Basic knowledge of Laos). Mekong Publishing, Tokyo (in Japanese)

Epilogue

The Implications for the Study on Population Dynamics and Livelihood Changes in Small-Scale Communities in Laos

Satoshi Yokoyama, Takahito Niwa, and Hongwei Jiang

Abstract This study describes the research approach adopted to identify population dynamics and livelihood changes in small-scall communities in rural Laos, and what research results were obtained as a result of the study. A diet survey based on a two and a half years' food diary in a swidden agriculture community in central Laos provided a valuable record of residents' dietary habits in a rapidly developing frontier region of Southeast Asia. Furthermore, the time allocation survey focused on daily life, which is vulnerable to the influence of market economization, and clarified the subsistence structure. In a village near the Thai border in central Laos, a retrospective survey was conducted to clarify the population dynamics from the establishment of the village to the present day, and the population dynamics and livelihood changes in rural Laos were clarified in light of the impact of migrant workers to Thailand. In the area of mixed ethnic groups in northern Laos, we clarified the problems of livelihood change caused by internal resettlement from the highlands to the basin, based on the possibility of acquiring agricultural land and the actual situation of migrant workers. In each of these areas, the research perspective of household reproduction as it relates to population dynamics was incorporated to discuss future issues in research on small-scall communities.

Keywords Population dynamics · Livelihood changes · Reproduction · Diet survey · Time allocation survey · Laos

S. Yokoyama (✉)
Graduate School of Environmental Studies, Nagoya University, Nagoya, Japan
e-mail: yokoyama.satoshi.i4@f.mail.nagoya-u.ac.jp

T. Niwa
College of Human and Social Sciences, Kanazawa University, Kanazawa, Japan
e-mail: niwa-2@staff.kanazawa-u.ac.jp

H. Jiang
National Institute for Environmental Studies, Tsukuba, Japan
e-mail: jianghw@gmail.com

S. Yokoyama et al. (eds.), *Population Dynamics and Livelihood Changes of Small-Scale Societies in Laos*, International Perspectives in Geography 22,
https://doi.org/10.1007/978-981-96-2956-5_12

1 Conclusion from discussions about Livelihoods in a Swidden Agriculture Community

1.1 Overview of AN Hamlet Study

1.1.1 Diet Survey in AN Hamlet

The AN hamlet diet survey provided a valuable record of the dietary habits of residents in a frontier area of Southeast Asia that is undergoing rapid development. In recent years, the development of plantations and mines by foreign capital has brought about significant changes in the lifestyles and diets of local residents. However, there are almost no systematic records of the traditional lifestyles and diets of local residents. These records are important not only for ethnographic study, but also for improving the quality of life and healthcare of local residents, as well as for environmental conservation in the region. In other words, these records can deepen various stakeholders' understanding of local residents' lifestyles and diets, and provide important clues for formulating health policies and mitigating side effects of subsistence transition. In this sense, the AN hamlet diet survey can undoubtedly provide important materials for the formulation of development policies and ethnographic studies in mainland Southeast Asia.

In addition to dietary content, another point worth mentioning was the analysis of food acquisition through life history study. The food acquisition analysis was based on food diaries of 11 households for two and a half years. It analyzed the food acquisition strategies of three categories of households: (1) young couples without children or with children under 10, (2) households with children aged 10 and over, and (3) elderly couples whose children are independent. These analyses revealed the relationship between the age structure of household members and their food acquisition strategy, as well as the contribution of gender and age group to food acquisition in a rural settlement in mainland Southeast Asia where the commodity economy has not yet advanced. These findings may provide important clues for understanding the time allocation of swidden agriculturist, variations in fertility, and predicting future population changes in mainland Southeast Asia.

1.1.2 Time Allocation Survey in AN Hamlet

The time allocation survey focused on the time allocation of daily life, which is easily affected by the transition to a market-oriented economy. Similar to the diet survey, the time allocation survey clarified the subsistence structure of AN hamlet based on the subjects' time allocation of daily life. Then, based on the theory of land use intensification (Boserup 1965; Carlstein 1982; Chayanov 1966), the possible responses of local people to large-scale agricultural development were analyzed. In conclusion, we found that (1) development led by foreign capital and the government's migration policy have put a real strain on land resources, with the result that the AN hamlet has been forced to change its subsistence activities; (2) the intervention of market factors has changed not only the time (labor) allocation of daily life, but also the degree to which individual subsistence activities contribute to survival, due to increased market value.

Of particular interest, this study revealed the spatial distribution of daily activity and energy expenditure. Unlike typical time allocation studies, the AN hamlet time allocation survey used GPS loggers and accelerometers. Analysis of the data collected by these instruments clarified the spaces in which the study subjects performed their subsistence activities and the locations in which their energy was expended. These results are expected to contribute not only to the understanding of environmental use by residents, but also to the assessment of the risk of environmental infectious diseases such as malaria.

1.2 Contributions to Research Methods

Although the food diary method applied in the diet survey is a research method used in previous study (Hasada and Yamada 2017; Nonaka et al. 2014), a valuable aspect of this study is that the diet of 11 households was recorded over a period of two and a half years, from May 2015 to November 2017and analyzed the division of labor within households and the flow of food among households, an important strategy for ensuring household food security in terms of food acquisition and redistribution. A total of 13,934 meals were recorded, including menus, sources of ingredients, cooking methods, etc. To the authors' knowledge, such detailed records over such a long period of time are extremely rare. The authors' ingenuity in designing the record form may have contributed significantly to their ability to collect data over such a long period. For example, based on long-term field surveys, the authors coded mealtime categories, gainers, and place of food acquisition. This reduced the amount of content that informants had to write and allowed for longer surveys.

In addition, several new methods have been introduced in the time allocation survey. Unlike conventional methods of time allocation survey, the "recall method" allows us to capture the subjects' activities over time, and the introduction of GPS loggers and accelerometers has made it possible to assess the spatiotemporal distribution of human activities quantitatively or semi-quantitatively from space and energy expenditure, as well as time length. In particular, there are few methods that combine GPS loggers and accelerometer data to assess the spatial distribution of energy expenditure. Such methods are expected to be applied not only to time allocation surveys of daily life, but also to time geography research.

1.3 Positioning AN Hamlet Study in This Book

Three key terms come to mind first to explain the positioning of AN hamlet study in this book. The first is the swidden agricultural settlement, the second is the Xepon River, and the third is the development of Xepon District. In contrast to the other areas covered in this book, the physical barrier of the river, despite its proximity to large-scale agricultural development, has allowed the so-called traditional subsistence of swidden agriculture to continue, with little change in the market economization of subsistence and migration for work. In a sense, it is an expedient study area for evaluating the subsistence and daily life activities of the local people at the beginning of agricultural development in Xepon District and in mainland Southeast Asia. With the advantage of this study area, the AN hamlet survey group was able to document settlement life from before or around the beginning of development. In particular, these records can provide valuable information for predicting the impact of future development on the lives of local people and for elucidating the primary drivers of change in areas that have already undergone change.

Unlike the other study areas, the population of AN hamlet has not experienced a population decline due to migrant workers, but rather a rapid population increase due to the government's migration policy. In a sense, they are entering a situation where migrant workers may emerge due to the shortage of land resources in the village, as they did in the third and fourth generations in KH village (Takahashi et al. 2025). However, at least at present, the phenomenon of migrant workers has not yet been observed in AN hamlet. This can be explained to some extent by the shift from swidden agriculture to profitable land use. It is very likely that local residents with different environmental and cultural backgrounds will have different ways of confronting foreign-led development and population growth. AN hamlet is only at the beginning of change, and its evolution will need to be closely monitored in the future.

2 Conclusion from discussions about Population Dynamics and Livelihoods in a Paddy Farming Community in the Plain

2.1 Population Changes in the KH Village and Effects on Migrant Work in Bangkok

2.1.1 Population Changes Since the Founding of the Village

The first outcome of our survey on the KH village was the clarification of population changes from the founding of the village to the present, through a retrospective method (Takahashi et al. 2025). The results indicate that livelihood strategies have changed according to the generation. It is thought that the first immigrants built KH village in the 1920s. Assuming that it takes demographic reproduction a quarter-century to effect generational change, the fourth generation appears as of 2014. Population growth rate had been high, at over 2% per year at least since the end of World War II.

At first, the acquisition of rice paddies supported a high population growth rate after the village was established. In the second generation, not only the acquisition of rice paddies but also the inheritance and purchase of paddies came to play a role in supporting population growth. However, in the third and fourth generations, it became impossible to support population growth with limited paddies due to a lack of suitable uncultivated land, and residents started to move to other villages or to Bangkok for work. Rather than an attempt to support population growth through improving agricultural productivity (Boserup 1965), the result was an adjustment of the population itself through migration.

2.1.2 Effects on the Village of Migrant Work in Bangkok

The second outcome was a clarification of the relationship between the village and Bangkok through interviews with those who had experienced moving to Bangkok for work (Niwa et al. 2025b; Shirakawa et al. 2025). There is a strong relationship between migrant work in Bangkok and population changes in the village, which are changes related to both of natural change in birth behavior and social change.

Migrant work in Bangkok has influenced birth behavior in the form of the introduction of modern contraceptive methods such as pills and injections to the village by those with experience working in Bangkok, especially female workers. These women began using modern contraceptive methods before family planning programs were introduced by the Lao government. Primary reasons for the use of contraceptives are the difficulties of childbirth and childcare, and the economic situation. On the other hand, the context of the spread of modern contraceptive methods includes the important roles played by healthcare professionals and medical institutions in acquiring knowledge of contraceptive methods and initiating use. The Lao government's family planning program was broadly promoted starting in the 1990s. The Lao Women's Union played a central role in promoting this program. As a result, the birth rate in KH village began to fall in the 1990s. This corresponds to the period of expanded use of modern contraceptive methods. The population has started to decrease slightly since the 2000s.

One of the factors of these population shifts is the vast increase in people migrating to Bangkok for work. Villagers choose to migrate to Bangkok for work as the main way to acquire cash because employment opportunities are limited in KH village and the surrounding area, including Savannakhet. It is obvious that migrating for work is regarded as more important than education, according to the low level of educational attainment of migrants. We found that the strength of villagers' family networks and local networks, as well as the expanded opportunities to encounter information about Thailand, have made it easier to migrate to Bangkok. As a result, the residential population of the village is starting to decrease. In addition to this, the adjustment of the population size caused by migration was not brought about by government regulations or discussions held by the village. It seems that these results were caused by livelihood strategies at the level of each person and each household.

2.2 The Relative Positioning of KH Village

Comparing KH village to NK village, HB village, or AN hamlet, KH village has a feature that population migration to Thailand acts as a trigger for change in rural villages. In KH village, the short physical and psychological distance from Thailand seems to be linked to migration to Bangkok for work. In other words, the fact that almost no one from other villages studied migrated across the border may be influenced by the spatiality of their closeness to Thailand. The earlier KH village had acquired basic infrastructure such as roads and electricity than the other villages studied, which led to it likely that the village had many opportunities to receive information from Thailand.

Modern contraceptive methods did spread in the other villages studied, and it is confirmed that the birth rate has been dropping, however, in the case of KH village the residential population is decreasing due to greater migration to Thailand. Population changes are affected by both natural and social changes: when the reproductive-age generation moves away, as in KH village, the birth behavior of that generation also takes place outside the village, which in turn affects natural change. Additionally, some parents send their children back to KH village, forming "skipped generation households". The relationship between the village and this kind of birth behavior is strongly affected by the Thai government's policies toward foreigners. Therefore, population changes in KH village may be strongly swayed by influences external to the village. NK and HB villages are also experiencing lower birth rates and migration toward cities, but their populations are maintained through the influx of people from mountainous regions.

To summarize the above, our studies show that rural parts of Laos that are closer to cities, that is those in an environment where it is easier for people to migrate to cities, are more strongly affected by population decline in the twenty-first century. In developed countries, hinterlands far away from cities show more advanced rural depopulation (Ishikawa 2020; Johnson and Lichter 2019), but in Laos, this may be advancing in not only the highland ethnic minority villages in which there is no choice but to migrate to the lowlands due to policies that encourage migration, but also rural villages closer to cities, even in lowland villages with wet rice paddies, in addition to. While it is difficult to predict future trends, if the industrial and service sectors expand in Laos, rural villages like KH village may be unlikely to maintain their populations through rural–rural migration.

2.3 KH Village Study and Fieldwork

The survey conducted in KH village provided more long-term and detailed population data than the population census to understand of demographics in this area. The population was reconstructed through the retrospective survey, and the method itself has not changed significantly. However, this study collected complementarily related information through surveys grasping the situation of farmland ownership and work conditions at migrants' destinations. These allowed a more precise reconstruction of the population. Nevertheless, there may not have been enough analysis of the child-woman ratio or early childhood mortality due to issues in the population size of the study village.

Studies on rural development in developing countries often discuss rural surplus labor and industrialization based on the Lewis Turning Point (Bowonthumrongchai 2019; Yamada 2016; Lewis 1954; Ranis and Fei 1961). The case of KH village can be considered a topic for this discussion through the migration of its population toward cities, which has occurred in the third generation onward. Instead of moving to cities in Laos, villagers from KH village moved to Bangkok, Thailand. Taking this into consideration, while the rural–urban migration from KH village may have contributed to Thailand's industrial and economic development, it did not contribute to the capital accumulation in Laos. Although we confirmed large amounts of foreign remittances in our studies, these funds did not contribute to the accumulation but were invested in education expenses and farmland. Laos may not be able to catch up industrialization that other Asian developing countries have experienced (Suehiro 2008).

2.4 Effects of COVID-19

We conducted interviews with 67 village households in March 2023 and gained an understanding of household circumstances. The results confirmed that 110 household members had migrated to Thailand. When applied to the entire village through simple proportional distribution, the result would be 209 people. On the other hand, the residential population is 380. Considering the residential population base, there has been a 32% decrease in population from December 2014 to March 2023. Seen in relation to livelihoods in the village, the wet rice paddy area per capita in KH village may have increased due to population decline. In the context of even more migration to Thailand than before COVID-19, there may have been affected significantly on household livelihoods when people were compelled to return to the village due to COVID-19 (Suhardiman et al. 2021). When they were able to cross the border again, there may have been a reactionary expansion of movement toward migrant work. Additionally, there has been a large increase in the number of villagers who seem to have migrated to Thailand permanently instead of temporarily. Villagers from KH village mentioned that many migrants married Thai nationals. The greater number of marriages with Thai may have been due to the actions taken by villagers who experienced the emergency of COVID-19 while abroad, in order to reduce the risks of staying and living abroad. Since this point has not been clear, it remains a topic for future research. Detailed surveys and analyses are needed to clarify the current state of the village, the factors that strengthen its ties with Bangkok, and whether the migrants in Bangkok are maintaining relationships with KH village.

3 Conclusion from discussions about Population Dynamics and Livelihoods in Both Paddy and Swidden Farming Communities in the Basin

3.1 Internal Resettlement in Northern Laos

This study focused on the Muang Ngoi area, Ngoi District, Luang Phabang Province, in which the principal cause of changes to population dynamics and livelihood change was internal resettlement by both voluntary and involuntary migration in order to seek out better livelihood infrastructure. Accordingly, this section discusses internal resettlement, which we were unable to take into full consideration in Yokoyama et al. (2025), Niwa et al. (2025a), and Niwa and Nishimoto (2025).

3.1.1 Three Internal Resettlement Programs in Laos

Internal resettlement in Laos is directly related to three important programs: "Focal Sites," "Land and Forest Allocation," and "Village Consolidation" (Baird and Shoemaker 2007).

Focal Sites is a program that has been running since the early 1990s. It concentrates highland minorities in selected areas, reduces poverty among remote rural residents, provides food security, promotes the commercialization of agricultural products, and improves access to development services (Government of Lao PDR 1998). In 1998, the Lao government announced that it would establish 87 Focal Sites by 2002; it was resettled 1,200 villages and 450,000 people (12% of the rural population of Laos) (Évrard and Goudineau 2004). The government guided highlanders to Focal Sites because it would cost less to provide services to highlanders in lowland areas than to provide them with services and supporting infrastructure where they were. This was for economic reasons and was accomplished with large donations from international aid agencies.

The Lao government also worked on substantial land reforms as a major initiative starting in the early 1990s. The focal points were reinforcing security of land tenure to encourage farmers to participate in intensive farming and discouraging the practice of swidden agriculture in order to protect the environment. The program in which forests were classified according to their uses and places classified as farmland were distributed to residents was called "Land and Forest Allocation". It was introduced provisionally in the northern provinces of Luang Phabang and Sayabouly in 1990 and implemented nationwide in 1994 (Ducourtieux et al. 2005). The forests that local residents had been using until then came under control of the district office, and the inability to practice swidden agriculture encouraged people to resettle to the lowlands.

"Village Consolidation" was a way to make government management easier by merging small villages scattered in the highlands. A single village size was decided in the Decree on the organization and administration of villages (Prime Minister's Office No. 102/PM) issued in 1993, with a village required to have at least 20 households or 100 people. Villages with fewer than 20 households would be unable to become independent villages and were placed under adjacent villages. Afterward, Article 51 of the Law on Local Administration, revised in 2003, stated that there would no longer be any requirement for the number of households related to establishing new villages. Instead, there would be minimum populations: 200 for highlands or remote areas, 500 for lowlands, and 1,500 for urban areas. This provision still stands today.

3.1.2 Village Consolidation in Luang Phabang Province and Decreasing Village Numbers

Since the study area was not designated as a Focal Site, most of the population movement from the highlands to NK and HB villages is due to voluntary resettlement or the Village Consolidation policy.

The average village size calculated from the population and number of villages in Luang Phabang Province is shown in Fig. 1. In the 15 years from 1985 to 1999, the number of villages remained consistent at roughly 1,200. However, this number dropped from 1,162 in 1999 to 951 in 2000, a decrease of 211 villages. As the population continued to grow from 2000 to 2015, the population per village increased rapidly. It was over 400 people per village in 2000, about 500 per village in 2005, and over 600 per village in 2015, showing that village size had been growing.

However, village population growth does not necessarily indicate that the population is geographically concentrated in one place. Yokoyama (2004) conducted a survey of the Pak Luang area of Ngoi District in 2001 and wrote that while there were 21 villages in that area in the mid-1990s, the number had decreased to 14 villages as of 2001; the reason was that Village Consolidation had been conducted by the district administrative office. At that time, there was no actual resettlement of residents, and the population of the smaller villages was simply consolidated under the larger neighboring villages.

The purpose of Village Consolidation is to efficiently provide residents with services, such as education and sanitation, with emphasis placed on improving access to the market. For this reason, these services are not provided to highland settlements with fewer than 20 households that fall under the administration of another village. As Fig. 1 shows, the number of villages fell as Village Consolidation advanced rapidly from the late 1990s to the mid-2000s. Residents of small highland settlements that

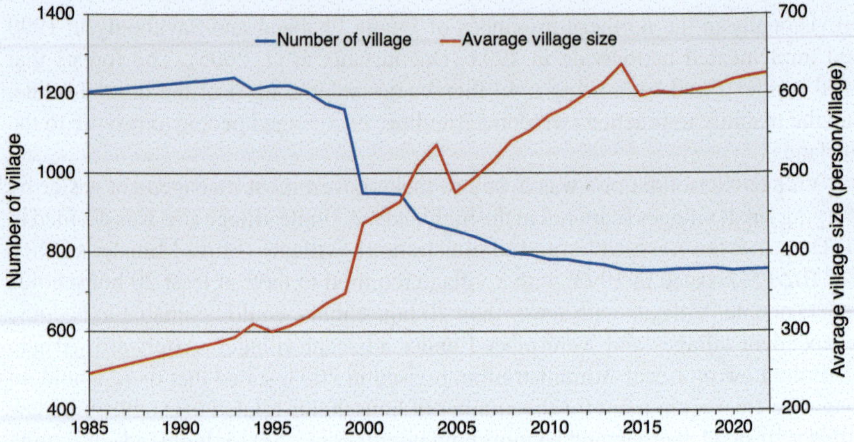

Fig. 1 Number of villages and average village size in Luang Phabang province, Laos. *Source* Data from Statistical Year Book, National Statistical Center, Laos

were consolidated into other settlements had to decide whether to keep living where they were or resettle in search of public services and new employment opportunities.

3.1.3 Resettlement Issues in NK and HB Villages in Ngoi District

NK and HB villages in Ngoi District have received many incoming residents from highlands. Table 1 shows the number of people from highlands in Ngoi District who resettled in NK and HB villages, the lowland receiving villages. This table excludes data on moves due to marriage and moves from neighboring villages. Furthermore, moves made by people from outside Ngoi District were not included because their reasons for whole-household migration were different. However, the data in Table 1 shows that both villages received large numbers of migrants during the Lao civil war period of 1953–1975. Unlike the Village Consolidation, the migration during this period was an emergency migration. A total of 17 villages were on the "sending" side in Ngoi District. Of these, as of 2015, 11 were villages that had been merged with neighboring villages and lost their names, or villages that were abandoned completely. It seems clear that nearly every year, both villages continue to receive residents from small, isolated settlements in the highlands due to Village Consolidation. In the villages that these residents come from, people cannot receive government services such as education, sanitation and road access. One of the main factors in deciding to resettle may have been the extreme difficulty of continuing to live in such places.

However, incoming residents from highland settlements do not see great improvements to their lifestyle in their new villages. Since there is already no room left to reclaim rice paddies in NK village and HB village, new residents must continue practicing swidden agriculture because they are still unable to cultivate in paddy

Table 1 Migrants received in NK and HB villages and origin villages of migrants in Ngoi district, Luang Phabang province

Village status[1)]	Origin village of migrants[2)]	Destination village of migrants		Period of migration
		No. of migrants in NK village	No. of migrants in HB village	
Existing	Don	4	3	Immigrated in 1974, 1989, 2013 and 2018
	Kiukhan	4	2	Immigrated in 1992, 2002 and 2012
	Longkham	0	2	Immigrated in 1991 and 2015
	Phon	35	4	Immigrated in 1974, 1984 and 2018
	Sopchaem	3	1	Immigrated in 1977 and 2014
	Xang	49	2	Immigrated in 2004, 2005 and 2018
Not existing	Bouamaen	24	4	Immigrated in 1977, 1988, 2000, 2002, 2004 and 2015
	Khonkeung	3	0	Immigrated in 1974
	Kiuhia	21	6	Immigrated in 1974, 1981 and 2000–03
	Mokkou	21	0	Immigrated in 1974–75
	Nalo	5	0	Immigrated in 1970
	Nambong	10	14	Immigrated in 1957, 1976, 2002, 2007, 2013–14 and 2017
	Nong-ian	10	0	Immigrated in 2005 and 2014
	Saengsavang	9	12	Immigrated in 1991, 2003–05 and 2018
	Sen	0	5	Immigrated in 1972
	Sensoukha	0	4	Immigrated in 1974, 1990
	Sopkhan	0	7	Immigrated in 1954, 1996, 2000, 2003

Source Authors' interview survey
Note (1) Village names were searched using data from the population and housing census 2015 (Ministry of Planning and Investment, Lao Statistics Bureau). Existing" refers to villages found on the list in 2015, and "Not existing" refers to villages that were not found. (2) Migration between NK and HB villages, from neighboring villages, and due to marriages were excluded

fields. Previous studies have reported on the rising population density and increased difficulty of accessing farmland in migration destinations (Nakatsuji 2005). While providing services in a designated place for resettlement allows more people to enjoy these benefits, the problem arises that it becomes more difficult to access finite plots of land. This study found that the government objective of reducing swidden agriculture is not being achieved through resettlement.

Additionally, the study areas of NK and HB villages were originally ethnic Lao villages; when Khmu people from the highlands moved in, they became ethnically mixed villages. In other parts of Laos, there have been instances of issues arising between residents on the receiving side and incoming migrants over land ownership (Alton and Rattanavong 2004; Kramp et al. 2022). However, the current study areas have fortunately been spared any ethnic conflicts over land as of the present point in time.

Migration from highland to lowland villages, that is, migration between rural villages, may have lower psychological barriers than migration to urban areas because in both cases the principal occupation is agriculture. However, when a certain amount of time has passed after migrating, children go on to secondary schooling at urban educational institutions or find increased opportunities to migrate to work in urban areas; ultimately, there are many cases in which incoming migrants who do not own farmland move to the cities where their children reside.

When still living in the highland settlements, surely many migrants dreamed of acquiring farmland in their new place of residence, producing the rice they needed, and being able to live without any family members going away for work. However, the reality is that virtually no migrants have been able to realize that dream.

3.2 Responses to Population Growth and Reproduction

Due to the topographical characteristics of the basin area that was the target of this study, the area around NK village and HB village has limited land suitable for reclaiming rice paddies. Nevertheless, the resettlement of Khmu people from highland communities has resulted in population growth, leading to a situation in which most incoming residents in both villages are unable to own rice paddies. For these migrants in particular, life in either village is not easy in terms of economics, and hardship increases further as the number of children rises.

In this situation, the women of the study area were asked how they deal with population growth; as shown in Shirakawa and Moriki (2025), the results obtained in this survey of NK village were interesting. The Khmu people residing in NK village, who migrated there from highland settlements, clearly have a lower rate of modern contraceptive use such as the pill or injections, in comparison with the ethnic Lao. The women of NK village get information from a health center in NG village, the center of the district, and use modern contraceptive methods. The results of the survey implied that there may be an access gap among migrants regarding this information.

On the other hand, as the spread of modern contraceptives and migration to urban areas accelerates, population decline in rural areas is expected to accelerate in the future. A decrease in young people may especially generate labor shortage issues in rural areas of Laos where the livelihood structure relies on rice cultivation. It is only a matter of time for the population decreases seen in rural Thailand to become apparent in rural Laos as well (Rambo 2017). In rural Thailand, migrant workers from the CLM countries (Cambodia, Laos, Myanmar) make up for the labor shortage (Rungmanee 2016; Pholphirul 2012), but there is no equivalent to this to be found in rural Laos. This study should be able to provide important primary data for the formulation of policies concerning resettlement initiatives, migration of populations from rural to urban areas, and reproductive health.

Acknowledgements This research project was conducted with the cooperation of the residents of KH village, AN hamlet, NK village, and HB village, and the staff of the counterparts at the Ministry of Agriculture and Forestry, the National Agriculture and Forestry Research Institute of Laos, the Ministry of Health, and the Tropical and Public Health Institute, to whom the authors are most grateful. This study was supported from JSPS Grants-in-Aid JP25257004 and JP17H01633.

References

Alton C, Rattanavong H (2004) Service delivery and resettlement: options for development planning. UNDP/ECHO, Vientiane

Baird IG, Shoemaker B (2007) Unsettling experiences: internal resettlement and international aid agencies in Laos. Dev Chang 38(5):865–888. https://doi.org/10.1111/j.1467-7660.2007.00437.x

Boserup E (1965) The conditions of agricultural growth: the economics of agrarian change under population pressure. George Allen & Unwin Ltd, London

Bowonthumrongchai T (2019) Lewis turning point and agricultural development in the Mekong: a case of Thailand. Shakaikagaku Ronshū (Soc Sci Rev) 156:31–49. https://doi.org/10.24561/00018613

Carlstein T (1982) Time resources, society and ecology: On the capacity for human interaction in space and time. Vol 1. Preindustrial society. George Allen & Unwin Ltd, London

Chayanov AV (1966) The theory of peasant economy. For the American economic association translation series. Originally published in Russian in 1925. Richard D. Irwin Inc, Homewood

Ducourtieux O, Laffort JR, Sacklokham S (2005) Land policy and farming practices in Laos. Dev Chang 36(3):499–526. https://doi.org/10.1111/j.0012-155X.2005.00421.x

Évrard O, Goudineau Y (2004) Planned resettlement, unexpected migrations and cultural trauma in Laos. Dev Chang 35(5):937–962. https://doi.org/10.1111/j.1467-7660.2004.00387.x

Government of Lao PDR (1998) The rural development programme 1998–2002: The "Focal Site" strategy, Outline of an approach, Programme support and investment requirements. Sixth round table follow-up meeting. State Planning Committee, Government of Lao PDR, Vientiane

Hasada K, Yamada R (2017) Raosu chubu ni okeru shokuryou nyusyu no genjou to kadai (Current status and issue on food access in a semi-mountainous village of Central Laos. Water, Land Environ Eng (Suido no chi) 85(5):469–474. (in Japanese). https://doi.org/10.11408/jjsidre.85.5_469

Ishikawa Y (2020) Internal migration in Japan. In: Bell M, Bernard A, Charles-Edwards E, Zhu Y (eds) Internal migration in the countries of Asia: a cross-national comparison. Springer, Cham, pp 113–136

Johnson KM, Lichter DT (2019) Rural depopulation: growth and decline processes over the past century. Rural Sociol 84(1):3–27. https://doi.org/10.1111/ruso.12266

Kramp J, Suhardiman D, Keovilignavong O (2022) (Un)making the upland: resettlement, rubber and land use planning in Namai village Laos. J Peasant Stud 49(1):78–100. https://doi.org/10.1080/03066150.2020.1762179

Lewis WA (1954) Economic development with unlimited supplies of labour. Manch Sch 22(2):139–191. https://doi.org/10.1111/j.1467-9957.1954.tb00021.x

Nakatsuji S (2005) Diversification of livelihood activities in the hills of northern Laos: a case study of an administrative village along a main road. Geogr Rev Jpn 78:688–709. https://doi.org/10.4157/grj.78.688 (in Japanese)

Niwa T, Nishimoto F (2025) Rural-urban migration in northern Laos: a case study of HB village in Luang Phabang Province. In: Yokoyama S, Niwa T, Jiang H (eds), Population dynamics and livelihood changes of small-scale societies in Laos. Springer, Singapore, pp 191–206. https://doi.org/10.1007/978-981-96-2956-5_9

Niwa T, Nishimoto F, Takahashi S, Yokoyama S (2025a) Rural-rural migration in a rural village, Luang Phabang Province, northern Laos: focusing both on immigration and emigration in HB village. In: Yokoyama S, Niwa T, Jiang H (eds), Population dynamics and livelihood changes of small-scale societies in Laos. Springer, Singapore, pp 171–189. https://doi.org/10.1007/978-981-96-2956-5_8

Niwa T, Takahashi S, Nishimoto F (2025b) Transnational labor migration from rural village in Laos to Bangkok, Thailand: focusing on relationship with the home village. In: Yokoyama S, Niwa T, Jiang H (eds), Population dynamics and livelihood changes of small-scale societies in Laos. Springer, Singapore, pp 99–116. https://doi.org/10.1007/978-981-96-2956-5_5

Nonaka K, Ono E, Natsuhara K, Murayama N (2014) Shizen wo torikomu: nisshiho ni yoru Laos 3 chiiki no yasei shokumotsu sesshu no hikaku [Taking in nature: comparing wild food ingestion in three regions of Laos through the diary method]. In: Ikeguchi A, Sato R (eds) Nature and science kenkyu dai-3-kan: Shintai to seizon no bunka seitai [Nature and science research, vol 3. Cultural ecology of the body and survival]. Kaiseisha, Shiga, pp 61–82. (in Japanese)

Pholphirul P (2012) Labour migration and the economic sustainability in Thailand. J Curr Southeast Asian Aff Stud 31(3):59–83. https://doi.org/10.1177/186810341203100303

Rambo AT (2017) The agrarian transformation in northeastern Thailand: a review of recent research. Southeast Asian Stud 6:211–245. https://doi.org/10.20495/seas.6.2_211

Ranis G, Fei JCH (1961) A theory of economic development. Am Econ Rev 51(4):533–565. https://www.jstor.org/stable/1812785

Rungmanee S (2016) Unravelling the dynamics of border crossing and rural-to-rural-to-urban mobility in the northeastern Thai-Lao borderlands. Popul Space Place 22:693–704. https://doi.org/10.1002/psp.1989

Shirakawa C, Moriki Y (2025) Aspects of modern contraceptive use among women in rural northern Laos. In: Yokoyama S, Niwa T, Jiang H (eds), Population dynamics and livelihood changes of small-scale societies in Laos. Springer, Singapore, pp 207–222. https://doi.org/10.1007/978-981-96-2956-5_10

Shirakawa C, Nishimoto F, Takahashi S, Yokoyama S (2025) Growth in the use of modern contraceptive methods and the related social context in a central Laotian farming village. In: Yokoyama S, Niwa T, Jiang H (eds), Population dynamics and livelihood changes of small-scale societies in Laos. Springer, Singapore, pp 117–137. https://doi.org/10.1007/978-981-96-2956-5_6

Suehiro A (2008) Catch-up industrialization: the trajectory and prospects of East Asian economies. NUS Press, Chiang Mai

Suhardiman D, Rigg J, Bandur M, Marschke M, Miller MA, Pheuangsavanh N, Sayatham M, Taylor D (2021) On the Coattails of globalization: migration, migrants and COVID-19 in Asia. J Ethn Migr Stud 47(1):88–109. https://doi.org/10.1080/1369183X.2020.1844561

Takahashi S, Yokoyama S, Nishimoto F, Niwa T (2025) Population growth and generational response to livelihoods in a rain-fed paddy farming village of central Laos. In: Yokoyama S, Niwa T, Jiang

H (eds), Population dynamics and livelihood changes of small-scale societies in Laos. Springer, Singapore, pp 45–69. https://doi.org/10.1007/978-981-96-2956-5_3

Yamada T (2016) The turning point in the Vietnamese economy: revisiting labor surplus theory. Forum Int Dev Stud 47(5):1–15. https://doi.org/10.18999/forids.47.5

Yokoyama S (2004) Forest, ethnicity and settlement in the mountainous area of northern Laos. Southeast Asian Stud 42:132–156. https://doi.org/10.20495/tak.42.2_132

Yokoyama S, Takahashi S, Niwa T, Nishimoto F (2025) Population dynamics and paddy holdings in remote rural villages of northern Laos. In: Yokoyama S, Niwa T, Jiang H (eds), Population dynamics and livelihood changes of small-scale societies in Laos. Springer, Singapore, pp 141–170. https://doi.org/10.1007/978-981-96-2956-5_7

The manufacturer's authorised representative in the EU is Springer Nature Customer Service Centre GmbH, Europaplatz 3, 69115 Heidelberg, Germany. If you have any concerns regarding our products, please contact ProductSafety@springernature.com

Printed and bound by CPI Group (UK) Ltd, Croydon, CR0 4YY
01/10/2026
02225937-0004